Learning The English Wheel

William H. Longyard

Published by:
Wolfgang Publications Inc.
PO Box 223
Stillwater, MN 55082
www.wolfpub.com

Legals

First published in 2014 by Wolfgang Publications Inc.,
PO Box 223, Stillwater MN 55082

© William H. Longyard, 2014

All rights reserved. With the exception of quoting brief passages for the purposes of review no part of this publication may be reproduced without prior written permission from the publisher.

The information in this book is true and complete to the best of our knowledge. All recommendations are made without any guarantee on the part of the author or publisher, who also disclaim any liability incurred in connection with the use of this data or specific details.

We recognize that some words, model names and designations, for example, mentioned herein are the property of the trademark holder. We use them for identification purposes only. This is not an official publication.

ISBN-13: 978-1-935828-89-1

Printed and bound in U.S.A.

Learning The English Wheel

Page 10

Chapter One
Meet The English Wheel   *Page 6*

Chapter Two
What is the Right
E-Wheel for You   *Page 10*

Chapter Three
Getting Started   *Page 24*

Chapter Four
Tracking Patterns   *Page 30*

Chapter Five
An Aston Martin
Apprentice Panel   *Page 46*

Page 89

Chapter Six
Building Your Buck   *Page 52*

Chapter Seven
Birdcage Restoration   *Page 64*

Chapter Eight
Jamie Downie:
Back to The Future   *Page 84*

Chapter Nine
Make A Skylark Whole Again   *Page 96*

Page 126

Chapter Ten
Formula II Cooper Repair   *Page 112*

Chapter Eleven
Randy Ferguson   *Page 118*

Chapter Twelve
Kent White   *Page 134*

Sources/Education   *Page 142*

Dedication

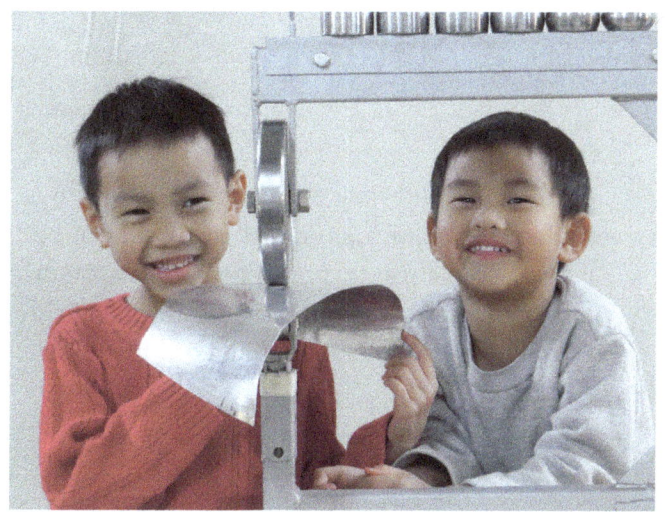

To my two favorite panel beaters, Huan Duy and Le Tien Bui who always say,
"We're the kind of people who make things, not break things."

Acknowledgements

How many times in your life do you actually get to meet the heroes of your youth? While researching this book I was lucky enough to finally meet and spend time with the great Ron Fournier, for which I will forever be thankful. More than just knowledgeable, Ron was a gracious and kind teacher from whom I learned as much in person over the course of several days as I had in 30 years of reading his books. I also got to speak at length to the venerable John Glover, the man who more than any other introduced America to English wheeling 40 years ago. Classic car restoration the world over owes you much, sir.

The craftsmen with whom I worked to tell the story of the English wheel were all blessed with great talents, but also great humility. They told me again and again that what they knew was only because of the generosity of their teachers. Not one of them adopted the condescending air of an artist, though from judging their work I knew each could not be faulted for at least some pride.

Again and again in researching this book, and meeting these craftsmen, I wondered if at Pebble Beach it ought to be the car owners who get the trophies, or more deservedly the panel beaters who revive, and sometimes even completely recreate, the rolling jewelry that delights the judges.

So thanks to all the wheelmen who contributed projects to this book. Most of you will appear in the next book in this series and you'll be joined by other talented metal shapers as well.

Chapter One

Meet the English Wheel

Brief History - Or - What's in a Name?

There is no such thing as an "English Wheel." Surprise! When these machines first began to be used in industry in England around 1900 they were called "wheeling machines." Their introduction coincided with the advent of automobile production, and they were originally used to smooth, or planish, crudely hammered out body panels. To planish means to turn a bumpy surface into a level plane. Early images of these machines show them with a crank handle on the upper

One of England's best panel beaters, Geoff Moss, rolls a Lotus panel on a historic Ranalah "wheeling machine" formerly owned by his friend the late Len Pritchard, co-founder of the renowned London coachbuilders Williams and Pritchard, Ltd.

wheel shaft. An operator pushed and pulled the handle while fellow workers fed a panel through thus flattening the bumps, or what metalshapers today call "walnuts."

According to car builder/historian Timothy Barton, who has done more work investigating the early history of these machines than anyone else, it wasn't until the early 1930s that the "wheeling machine" began to be used on a regular basis to form body panels. It had taken a generation for the lads on the shop floors of England's factories to figure out that their wheeling machines could be used to put shape into panels by crushing the metal, and thus spreading it. By spreading the metal in small increments with each pass an operator could actually raise crown shapes in a panel. Soon, the term "wheeling machine" gave way to "raising machine." The crank handles disappeared and now their use to form automobile, and aircraft panels became widespread. Apprenticeship programs were developed to teach the newly discovered possibilities of the raising machine and no one on Earth became better at it than the English.

Barton does not claim that the English invented the raising machine. Oddly enough, even patent searches in the U.S. and Europe have failed to identify the originator. However, by the 1950s the first machines

By simplifying the design of the traditional cast iron wheeling machine John Glover made it affordable for the masses, which led to the coining of the term "English Wheel" in honor of this immigrant from Britain. Glover, more than anyone else, saved the dying craft back in the 1970s.

Glover's good friend, and ace fabricator, Ron Fournier teaches students using the third of Glover's historic, original tubular steel e-wheels. Fournier himself deserves much credit for saving the art of metalshaping with his ground breaking 1982 book Metal Fabricator's Handbook. If you don't have it, get it.

had arrived in America. It wasn't until the late 1970s and early 1980s that knowledge of them trickled out to a wider audience of professional metalshapers, and amateur craftsmen. Most of the credit for that dissemination belongs to John Glover. Glover began wheeling as a 14 year old apprentice in England during WWII. In the lean post-war years he immigrated to Canada where he plied his craft for deHavilland, the aircraft builders, before then being hired away by General Motors in Michigan to build prototypes and special projects.

In Detroit in the 1960s Glover had access to a few crude home-made wheeling machines that colleagues in the custom metal shaping trade owned, but they worked poorly. His need for a good machine increased as his reputation for custom fabrication spread and he gained more and more after-hours work until finally, in 1976, he decided to design and build his own machine. It took him three shots to get it right, but what resulted was a very lightweight, mobile machine that could form as precise a panel as the best of the cast British-built wheels back in Blighty.

Glover had been forming panels for a fellow GM employee who was also a member of the Experimental Aircraft Association. This man, Jerry Williams, suggested that Glover draw up plans for his wheeling machine which were then featured in an EAA publication. The floodgates opened and Glover was swamped with requests for plan sets. Cal and Joyce Davis, the owners of the tool retailer business Metalcraft, offered to fabricate and sell examples of the machine, and it was Cal (according to Glover's wife Mary) who coined the term "English wheel," in acknowledgement of Glover's national origin. A micro-industry of English wheel builders sprang up all across America, most building some version of Glover's design. The art of the "wheeling machine" was reborn with a vengeance, and just in time, too. The 1930s-40s generation of war-time factory metalworking craftsmen were then retiring and departing the scene taking with them all of the great secrets of their craft. Thanks to John Glover though, a new generation got the affordable design they needed just in time to save and revive the craft. Today, in England, throughout Britain, and across the world, the terms "wheeling machine" and "raising machine" are giving way to "English wheel." I guess that's not such a surprise after all.

THE WHEEL AS A ROTARY HAMMER

The English wheel is nothing but a rotary hammer. As the upper wheel rolls over the panel, it continually strikes a new area with hundreds of pounds of force. The metal is spread out from the contact area. Changing the lower anvil determines the direction and amount of spreading.

The Stanguellini workshop, Modena, Italy in the 1950s. The big machine seen behind the 750 twin cam under construction is a maglio, the standard shaping tool of Italian carrozzeria during the golden age of coachbuilding. Fairly expensive, imprecise, and dangerous to use, it nevertheless was employed by many continental builders. Note the small electric motor on top that drives it. Note also the oldest of all metalshaping tools after the hammer, the dished tree stump. A wire buck hangs in the upper left.

WHAT THE E-WHEEL CAN AND CAN'T DO

Yes, the English wheel is a remarkable tool, but it can't do everything. The original wheeling machines started out as simple planishers. Tim Barton reports that they were also used to crush welds where panels joined. They still do this well. Once the handles came off they could raise crowns, create reverses, and blend transition areas. To a very limited degree, English wheels can shrink metal. Planishing often involves some component of shrinking.

However, English wheels (let's call them e-wheels from now on, shall we?) are primarily a stretching tool, not a shrinking tool, so in that sense they have more in common with a maglio than a power hammer. Because of the practical impossibility of shrinking with one, the wheeler needs to carefully pre-plan a panel so that he/she doesn't stretch an area that needs to be kept tight. The Yoder power-hammer operator can change dies and shrink out any mistakes, but the wheeler has to get it right the first time, or resort to some other type of tool to fix it.

Finally, let's end this chapter by stating… you're not going to become the Tintoretto of tin bashing in a weekend. But, it doesn't take decades either. Give yourself a few months of dedicated weekends and you'll be surprised at the beautiful panels you can wheel out. This book will guide you.

The Stanguellini family maintains a private museum at their Modena FIAT dealership. The wire buck for a race car is suspended from the ceiling, and on the front fender hangs an aluminum panel that has just gone through the first stage of shaping with a maglio. In the old days it would then be subject to planishing. The English wheel shapes and planishes at the same time, leaving a mirror-like finish- not the "bag of walnuts" seen here.

Chapter Two

What's the Right E-Wheel for You?

Cast Iron versus Fabricated Steel Frames

One hundred years ago when wheeling machines first began to appear, they, like most machine tools, were cast from molten iron. Welding was in its infancy, both electric and gas, and so if you wanted a stiff frame it had to be cast. Therefore nearly all the machines from the golden age of panel beating were cast, and mostly of English origin.

In a factory setting a cast iron frame had the useful qualities of sturdiness and durability. They penalized their owners, however, because of their weight which made them difficult to reposition on the shop floor, and expensive to transport. Cast

The best e-wheel you can get in Britain and Europe is this classic design from the company that probably produced the first wheeling machines in Britain circa 1908. The 373 is Frost's largest, most stable, most capable model. Many anvils are available. Though industrial quality and highly prized by generations of owners, it is reasonably priced.

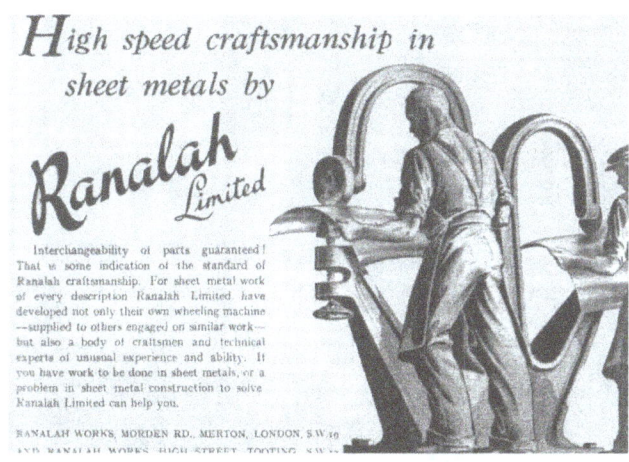

Amongst England's many wheel manufacturers Ranalahs were considered one of the best. Rugged, yet not over-weight, their design saved floor space due to their ability to be set back-to-back as seen in this ad from WWII.

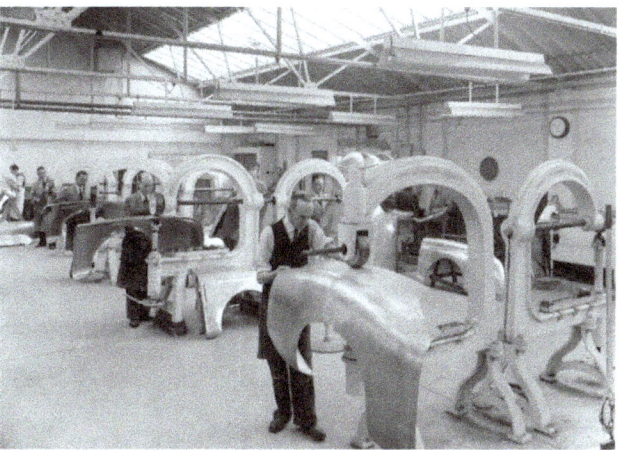

The Panel Craft shop 1951. In back they churn out Nash Healey fenders with Birmingham-built George Kendrick machines with lower adjusters, while the foreground operator uses a top adjustable Kendrick to crush gas welds.

machines might look bullet-proof, but in fact they did crack and were then nearly non-repairable due to the stresses imparted to the frame by the pressure generated between the wheels and the danger of the repair cracking again. If a brazed, or welded, repair gave way the operator could be seriously maimed by a falling chunk of cast iron, so a broken frame was usually scrapped.

The advent of welded tubular, and later fabricated steel plate, frames was a boon to metalshapers because these designs could be made much more cheaply than castings, and were also lighter, highly portable, and not subject to cracking. Because of this, there are many thousand more fabricated e-wheels in the world today than cast, even though cast wheels have been around much longer and are highly prized.

So why would anyone buy a cast wheel when, for much less money, they could have a fabricated wheel? The answer is that many believe that a cast wheel works better. They believe that because the frame is stiffer than most fabricated frames, the wheels will give a more consistent shape to the panel, and will actually work faster to raise the shape. They are correct… sometimes.

"Cast vs. Fabricated" e-wheels is one of those eternally debated questions like how many angels can fit on the head of pin, or what came first, the chicken or the egg, or blondes or brunettes? It may be fun to discuss for the philosophers amongst us, but I suggest a better use of time is picking one machine and getting on with learning how to use it.

Another Wray Schelin welded-plate machine in the making. Fabricated machines can be every bit as good as a cast machine when properly designed. Schelin comes up with machines that not only look cool, but work extremely well.

I built this machine in two days using 2 x 4 steel tubing. Keeping the tubing square during welding is more of a challenge than you might suppose due to the shrinking that always occurs along a weld seam.

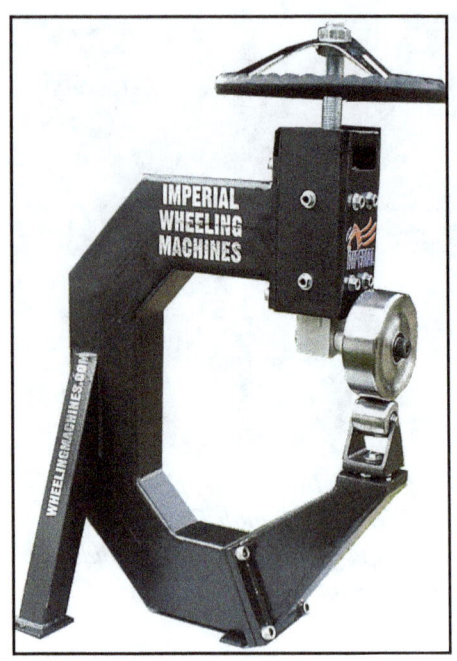
Owner of Imperial, Kerry Pinkerton offers wheels, adjusters, and yoke kits, and is a veritable fountain of information on all things English wheel. Great products and a great resource.

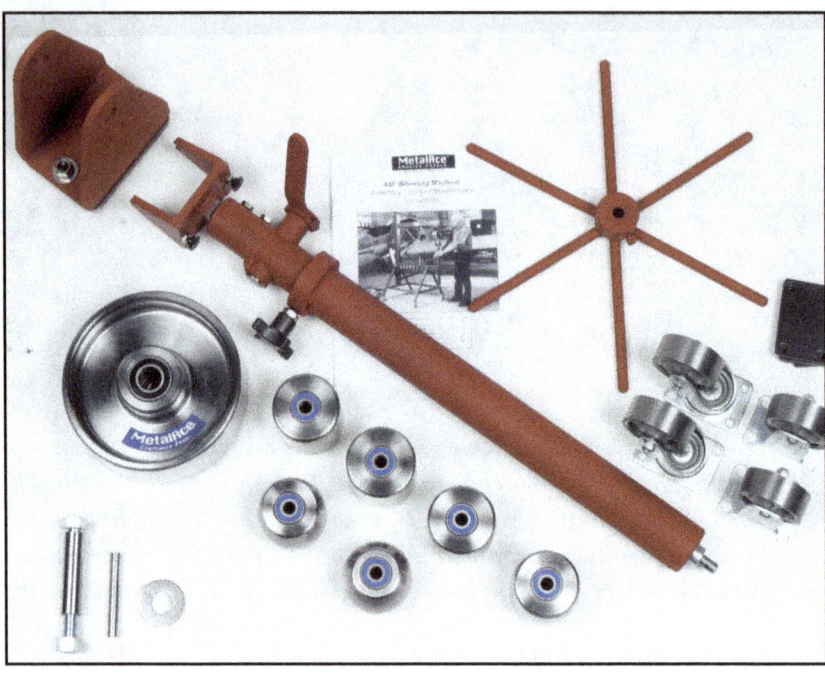
Metal Ace 44F U-Weld Kit: You can build a John Glover type machine with the quality components in this highly rated and popular kit. It comes with everything you need, including easy to follow blue prints. You supply the steel tubing.

I have used both and achieved good results with both. I have achieved bad results with some fabricated e-wheels, but that was because they were bad designs, and under-built. Either the tubing sizes were too small, or they were not sufficiently braced. Even bad designs can do a decent job planishing a panel, but the real test of a machine is consistent shape forming. Cast machines are highly likely to be quality shaping tools, while fabricated machines may, or may not, be. Test drive before you buy is the best advice on fabricated machines. Talk to others who have used the design before. John Glover's groundbreaking design looks spindly and frail, yet I have used it and it shapes as well as the best large cast wheels I've used. Glover understood that truss design could compensate for sheer weight, and so if you look at his machine you see lots of thin tubes triangulated to create rigid upper and lower arms for the wheels. Fabricated plate frames do away with triangulation and instead rely on "stressed skin" or "monocoque" construction. The welded panels form box beams that are every bit as good as a cast machine. Lazze Jensen's design is particularly noteworthy, being simple, lightweight, portable, and a dead-on reliable shaper. In Britain, Justin Baker's Raptor designs go a step further by drilling multiple lightening holes in the frame. These machines were the ultimate in strength and portability. Baker produces not only fabricated designs, but cast ones.

You can save an under-built tubular frame. I have had several that had springy frames or weak adjusting columns. One of them was so bad it had a hard time planishing .040 aluminum. Shaping was out of the question. I was able to save it by giving it a much stiffer vertical backbone and attaching a steel tube between its upper arm and a roof beam in my shop. It worked reasonably well after that, but was not portable, so I eventually junked it.

If I had my druthers, and didn't need to move my wheel around my shop as often as I do, I'd prefer a cast wheel. I really like their "Rock of Gibraltar" stability which makes guiding a panel through them somewhat easier, and even, to my mind, more intuitive. A cast wheel is dependable, visually pleasing, and it gives you a connection with all those great panel beaters of days gone by who built the most beautiful machines the world has ever seen. Such reasoning has more to do with romance than rationality perhaps, especially considering the fact that I use three fabricated wheels in my shop on a daily basis! One is a welded steel plate "C" frame Lazze-clone. It is very stiff, set up on rollers, and works extremely well. The second is a simple tubular design I built in two days. Being light and portable, I use it to take to demonstrations. It is the machine I set up when I need to use a go-kart slick for quick shaping (more about that later). The third is a little bench model that I much modified because the yoke support had too much play. I use its small 1 inch anvils to get into the tightest corners. I also let children use it when I'm conducting demonstrations at schools.

Spring Doesn't Mean Sprung

There is going to be discernible lifting up, or spring, of the upper arm on most e-wheels. That doesn't mean the design is poor. When you feed a panel between the upper wheel and anvil, you want the machine to exert hundreds of pounds of pressure to spread the metal. Remember, the machine creates a rolling series of hammer blows. In the same way that your arm is flexible, and will rebound upwards after striking with a hammer, the machine arms want to spread apart when under load and can be seen doing so. Look the next time you wheel. As long as the arms are sufficiently stiff to generate the pressure needed to shape, not just planish, the panel thickness you want to use, the machine is a good one. Let's be clear: ALL English wheels, even cast ones, must have some vertical flex (pushing apart up and down) between the wheels to work properly. What they should not have is any flex along the horizontal plane, or torqueing of the arms.

I wouldn't expect the best cast machine from the 1930s to form the steel plates used in ship construction, they would be too weak, and so I'd use something much bigger (they're out there!) Yet most automobile panels are going to be no thicker than .063, or 1.5mm aluminum, or 18 gauge steel, so it is pointless insisting on using a machine that was

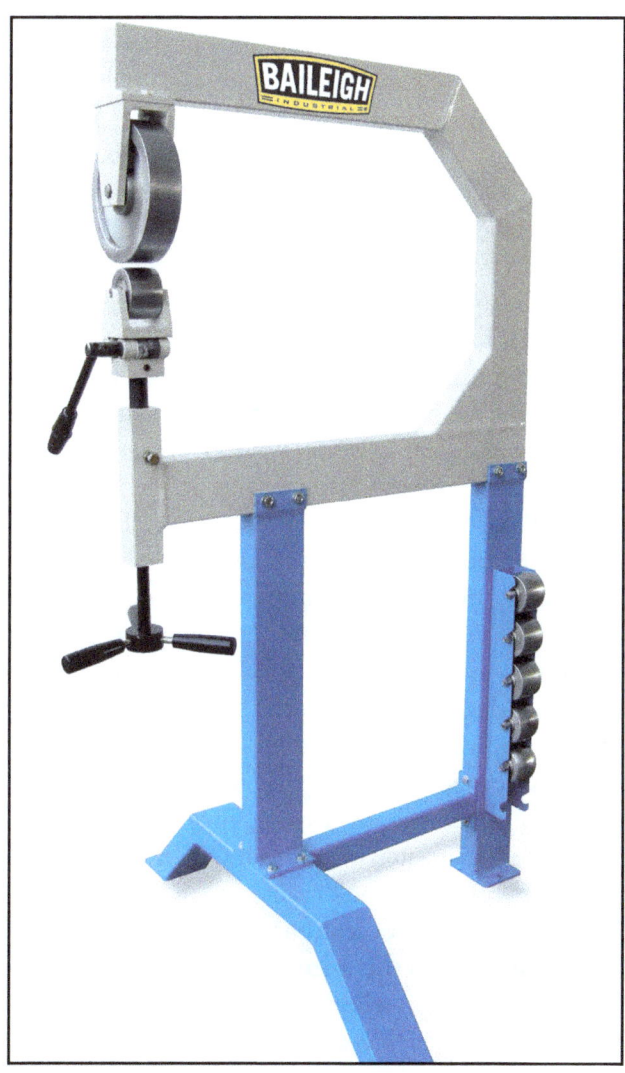

The low end of the e-wheel market is flooded with imported models of varying quality. This Baileigh EW-28, however, offers big improvements over look-alike competitors in yoke adjuster design, fit, finish, and anvil quality. Surprisingly capable, it can be demounted for use as a bench model. An excellent value.

This Metal Ace 22B Benchtop has been around for years, and got a lot of e-wheeling careers started. Its compact design is suitable for small shops - patch panels, and motorcycle parts. Made in US of A, its upper wheel and anvils are significantly better than most imported machines. Standard Duty

Metal Ace MA44F Floor Model: A classic which can roll panels with the best cast machines yet is easily moved around the shop, and can travel. Lots of notable American wheelmen have successfully used this design. A cost-effective professional U.S. made tool.

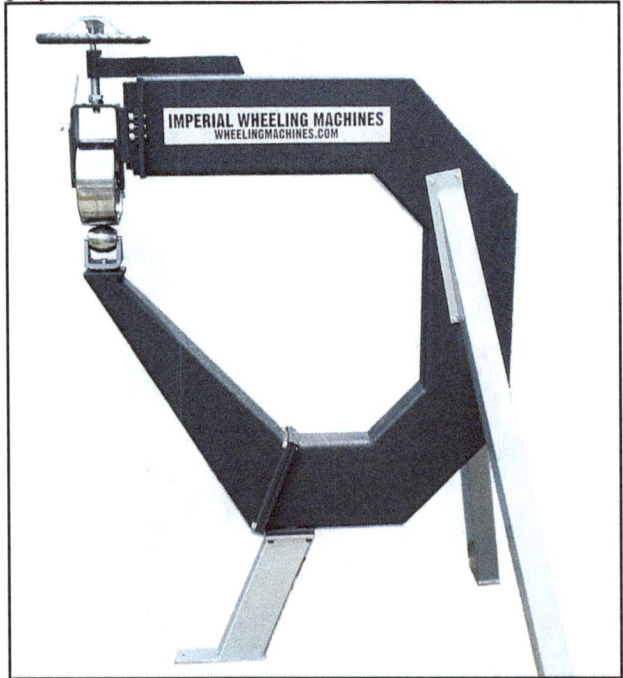

The largest of Pinkerton's excellent e-wheels, the Imperial Wheeling Machine Crown Imperial offers a quick release, and dovetail slide adjustment on the yoke. It stands level and unshakable on its unique tripod legs and is completely coated in Line-X to resist shop rash. Though custom made, it's affordable, and highly recommended.

designed for thicker material, like most cast iron machines were. If the wheels spread apart under load, but return to shape, in other words, they spring but haven't sprung - and if the part is being shaped consistently, there is nothing wrong with the design.

WHICH TO BUY?

The vast majority of shapers will find all requirements met with a fabricated e-wheel. The design should have a good reputation, whether store bought or built from plans, and there should be numerous craftsmen who are willing to vouch for it. The Metal Ace 44 is the lineal descendant of John Glover's original design, and as such has an impeccable pedigree. Matthys in Belgium sells a similar machine to the European market. Treat the frame you choose as a separate consideration from the anvils. Anvils are the real heart of any English wheel so make sure your frame comes with quality anvils, or simply buy the frame alone, and choose to purchase a quality set from another vendor.

A cast wheel is a bit of a luxury for a small shop, but probably worth it for a larger shop where cost and portability are less of a factor. When buying a used cast machine don't worry about worn bearings or bushings, as these are easily replaced. Missing anvils are easily replaceable even in the original pattern and axis diameter. Justin Baker in the UK makes them for all the classic English machines. Pay more attention to ensure the frame isn't cracked. Walk away from any that is suspect. Treat a new coat of paint with suspicion. A cast machine is a high-dollar investment so some spot wire-wheeling and a portable Magnaflux kit used on suspect areas could save you an expensive mistake.

SIZE MATTERS

Most novices to e-wheeling are like weekend yachtsmen. The latter tend to buy too much boat their first time out, and because of the hassle associated with its size, put it in the water less often than they planned, and therefore don't really enjoy it. The same is true with e-wheels. A machine that is too big only takes up valuable shop space, is hard to move, costs more, and often prevents you from doing smaller parts well.

Since 95% of all panels are less than 36 inches x 36 inches, and could therefore be rolled in a machine whose wheel to frame throat depth is only 18 inches, why do you need a large machine with a

48 inch frame? Even if you were rolling 48 inches x 48 inches you would only need a machine with a throat that's 24 inches wide. I've heard some experienced wheelmen admit that they could do just about anything on a machine with a throat depth of only 22 inches, especially if the machine featured upper and lower wheel carriers that swiveled 90°. In the old days occasionally entire fenders or quarter panels that were built up from multiple smaller panels would be run through an e-wheel to crush the welds, or level adjoining surfaces. Yes, a bigger machine would be required for such an operation, but realistically, that type of weld crushing and metal finishing can be done with a slapper and a dolly…though this is more time consuming. Bigger simply isn't better. I recommend spending less money on acquiring a larger tool, and more money on better anvils and upper wheels.

High or Low

The gap between the upper wheel and the lower anvil can be controlled by adjusting the height of either wheel. Some prefer to use an upper adjuster because that leaves more maneuvering room around the lower anvil when wheeling such high crowned panels as fenders. Being able to bring the panel around the underside of the lower anvil has advantages in some situations.

The more traditional placement of the adjuster is on the lower arm. A low-set adjuster allows the operator to keep both hands on his project while using his foot to nudge the tension higher or lower. Apples, oranges… I prefer the lower style because I've been in motorcycle, sports car, and even airplane crashes that have left me with torn rotator cuffs thus making it uncomfortable to reach up to a top mounted adjuster. I also don't like the idea of taking one hand off a panel because of the possibility of the

Kent White has a reputation for precision tooling, his 25.5 in. benchtop e-wheel is no exception. Every part, from bronze bushings, to precision ground acme threads, speaks quality and durability. Kent has established a loyal following amongst aircraft restorers, Ferraristas, and airline sheet metal shops. This pro workhorse can be fixed in place, or transported to the worksite. Industrial quality.

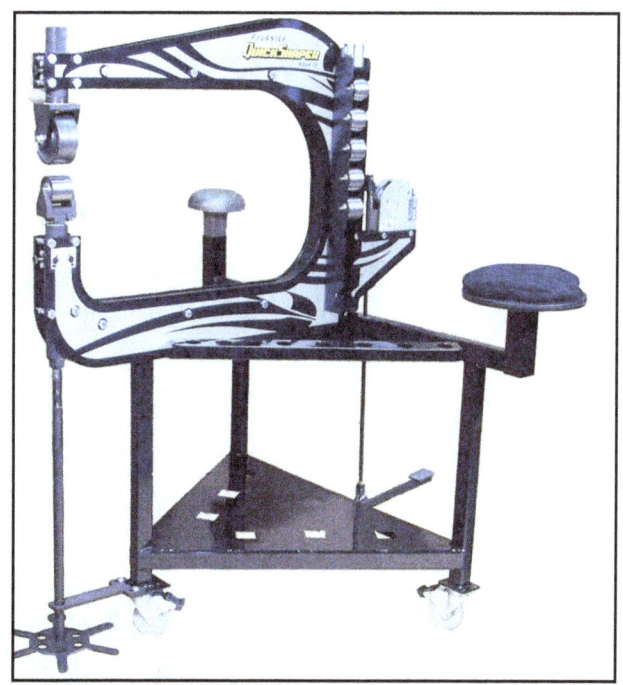

Fournier Quick Shaper Mk II: Leave it to Ron to design one of the most practical, space-saving work systems on the market. This multi-station work center allows shapers to have a portable e-wheel, forming stake holder, and shot bag in one location. The 2 in. anvils get into tight corners yet raise panels fast. The design is so good that Eckold builds a licensed version of it in Switzerland.

panel sliding between the wheels and then, due to its overhanging weight, getting a crease. The trend is towards upper adjusters, however, so it seems the argument about more maneuverability around the anvil yoke is winning out. Kerry Pinkerton's excellent designs make an eloquent case for this arrangement.

English Humour: I wasn't the first guy to think about using a roof beam to brace an e-wheel. In the very early days of wheeling machines in England, in the 1920s, some factories installed machines built in two components, a roof beam mounted upper wheel bracket that extended all the way down to just above an entirely separate floor mounted pedestal anvil holder. Factories trying to save money would buy several of these contraptions and set them up in a row directly under a beam. The row of operators soon found that if any one of them pulled a panel out from between the wheels, it affected all the others machines due to the immense pressure exerted by each machine on the common roof beam. It was a practical impossibility to keep the operators working in harmony and so the "money savings" idea was soon scrapped.

THE TRUTH ABOUT ANVILS: CONTINUOUS RADIUS OR FLAT

Another one of those apple vs. oranges, Ginger vs. Mary Ann debates is whether or not the anvils should have flat spots machined on them, or whether they should be of "continuous radius", no flat spots, just rounded. To understand this debate, you have to think of an anvil as a car tire.

A 7 inch wide tire does not touch the road across its full 7 inches. The "contact patch" may be more like 5 ½ inches wide, and instead of being a nice straight line side to side it will be more of an

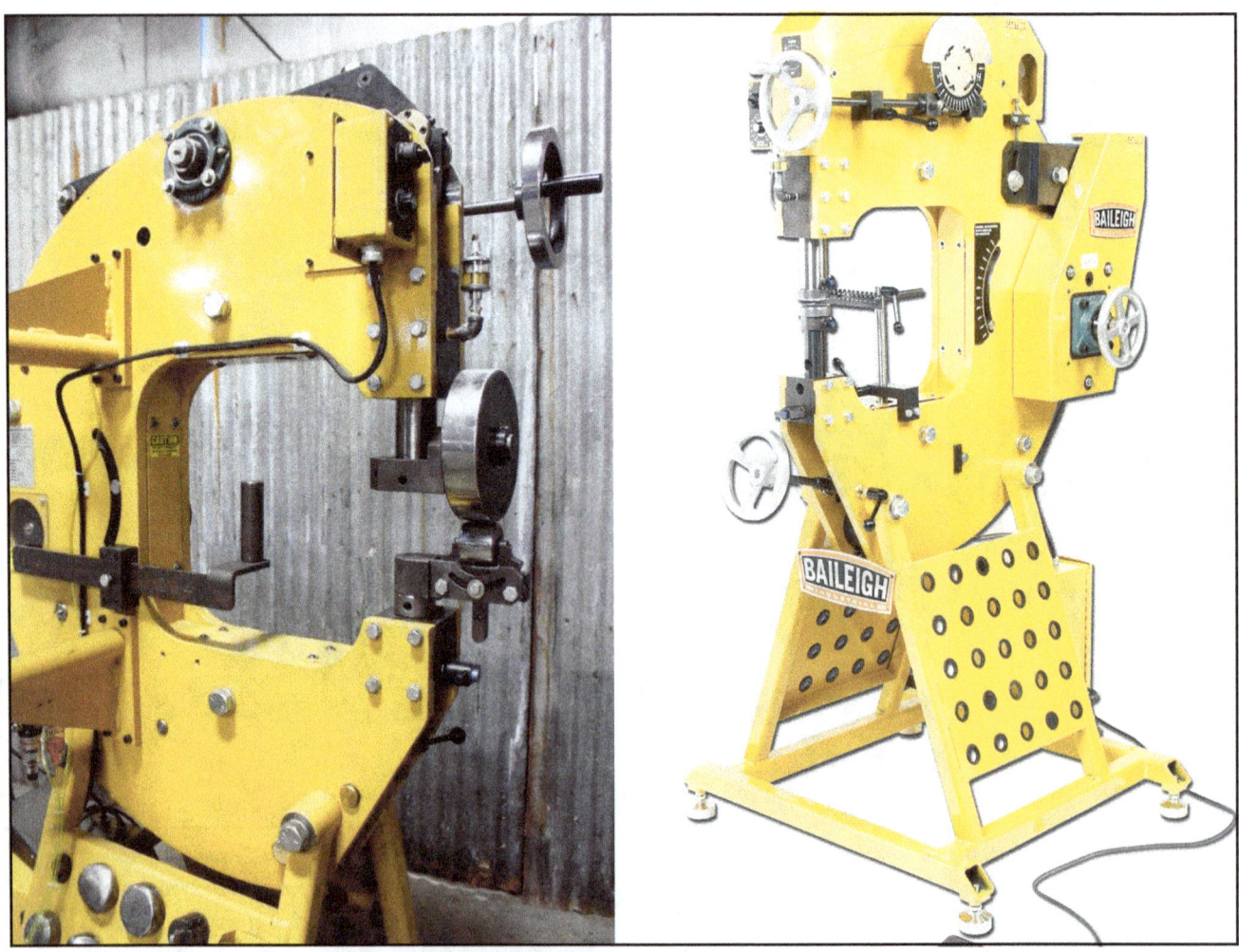

This is an utterly remarkable tool. Primarily designed as a power hammer (one that is much more user friendly than a Yoder), the MH 19 also quickly converts to an English wheel. I've used products from all the companies listed in this book, but none has more impressed me than the MH-19 due to its ease of use, shop-friendly foot print, portability, and transformability. Worth looking into if your shop is expanding and you need the speed of a power hammer.

oval. Anvils too create a contact patch with the metal above them. On a flat profile anvil the contact patch is a full width 2 inches on a 2 inch anvil. However, as the radius of the anvil decreases side to side the patch width rapidly decreases. Therefore, some anvils are cut so that the contact patch is set at a predetermined width. This diminishes the ridges and tracking marks as the panel is rolled back and forth.

Others prefer not to have a contact patch cut into the anvil, believing that the radius of the metal will naturally find its own best contact patch and fall away from the point of contact in a gentler curve, rather than the more abrupt transition left by the flat of a precut anvil. John Glover himself vehemently opposes this notion and calls a continuous radius anvil "dangerous." In fact, he doesn't consider a continuous radius anvil an anvil at all. Rather, he scornfully terms them a "fast raising wheel." When interviewed for this book he said, "I can't see how anyone can do any work with them, you can't." Glover's normal working set of anvils contained five flat cut anvils and one "fast raising wheel" which he only used in the tightest of corners and only with extremely light pressure.

Advocates of the continuous radius anvil say they are especially practical when working with a tubular steel framed e-wheel that has some spring in the arms. As the arms begin to flex away from each other during the insertion of the metal between the wheels, the axis of the upper wheel will no longer be parallel to the axis of the lower anvil. Thus, if using flat cut anvils, the flat of the upper wheel and the flat cut on the lower anvil will no longer be parallel. However, with a continuous radius anvil the slight offset from parallel will be automatically compensated for by the continuous radius of the anvil. This is absolutely true and indisputable. However, it is also indisputable that Glover, and many others using flat cut anvils in steel framed e-wheels have made perfect panels, repeatedly.

Baileigh EW-37HD: Who says the best e-wheels were designed in the 1930s? Baileigh broke new territory with this massive 37" throat, fully adjustable brute made from twin 5/8 inch thick steel plates. This machine can do everything from planish patch panels to crush steel welds on highly crowned fenders. Thick aluminum boat skins are no problem. An absolutely rock-solid machine that can even be dismantled for transport. Despite its size it is priced very reasonably.

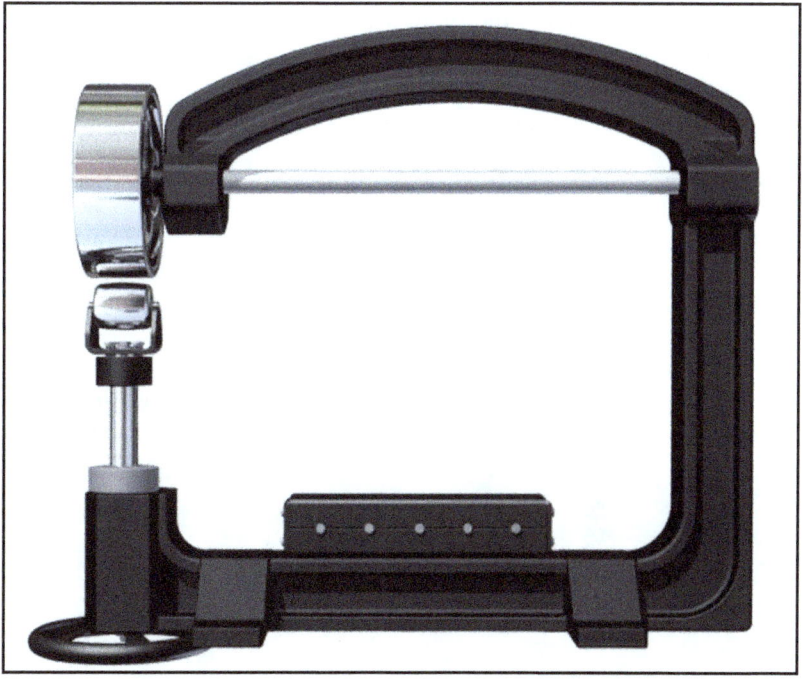

Justin Barker Traditional Benchtop: From the land that brought you the delightful original Austin Mini comes this wonderful mini cast e-wheel. Most shops simply need nothing more than this quality tool. Comes complete with a set of high quality Baker anvils.

Metal Ace-Imperial: The best e-wheels produced in the U.S. are Metal Ace's Imperial 28 and 44 inch models (not to be confused with Kerry Pinkerton's company). Cast in a modern foundry in Ohio, they are then precision finished in Iowa and feature Metal Ace's proven quick release mechanism, and quality anvils. A tool to be proud to own.

The locking bolt for the yoke column. In real world e-wheeling it must not be locked.

Continuous radius anvils are more difficult to use because they are utterly unforgiving. Your tracking pattern always has to be very narrow (more about this later) and you can only use very light pressure or you'll create furrows on the underside of the panels that are laborious to remove. Flat cut anvils obviate any of these worries. You also don't have to worry about ruining your upper wheel when withdrawing a panel too quickly as you must with continuous radius dies. With these the anvil can easily create a flat spot on the upper wheel if the two wheels bang shut when the panel is withdrawn. This is the main reason Glover deems them "dangerous."

Flat cut anvils are the most commonly used, and are what I prefer. However, I recommend spending less time worrying, and more time wheeling! A properly shaped panel is not finished just when the initial shape is put into it. The final step in wheeling all panels is to do one or two wash over passes. This is true whichever style anvil you use. A low pressure, closely tracked wash over is going to eliminate any residual shaping marks left by the anvils anyway, and leave the panel with a near mirror-like finish. So, take your pick… Ginger or Mary Ann.

UPPER WHEEL CONSIDERATIONS

The standard calculation for designing an upper wheel is that it be roughly three times the diameter of the lower anvil set. Thus, a three inch diameter lower anvil would ideally be matched to a 9 inch diameter upper wheel. Typical upper wheels for a common three inch diameter wheel set are eight to ten inches.

Many smaller e-wheels have two inch wide upper wheels. This width is perfectly serviceable, though a three inch wheel will actually give you a little more panel control and be a little more forgiving. The extra width helps stabilize the panel as it rolls through. A two inch upper requires the operator to be more careful about maintaining the panel's balance between the wheels and not letting overhanging weight pull twists in the panel, or worse, creases. This is not a theoretical concern, but one that happens all the time if you're unwary. For the slight difference in cost, I recommend going with a three, or better still, a four inch wide, eight or nine inch diameter wheel. Retrofitting an inexpensive introductory e-wheel with one of these, and a real quality set of anvils can transform a hobbyist machine into a real performer. I've seen machines with six inch upper wheels, but frankly think this is over kill for most applications.

ANVIL WIDTH

If you're doing larger panels, like door skins, fenders, and bonnets, you'll definitely want to go with three inch wide anvils. However, you don't have to, because you can get the same result with two inch anvils though you have to take more care, and make more passes. Most of the classic cast machines used three inch wide three inch diameter anvils.

The advantage of a narrower anvil is it allows you to get into tighter spaces, which is very useful. The advantage of wider anvils is they are marginally faster, and more forgiving. Quality trumps width, however. I have three complete sets of Hoosier Profile anvils that I use on my primary e-wheel. I use my three inch wheels 85% of the time, my two inch wheels 14%, and my one inch wheels 1%. I'm glad I invested in all three sets. My upper wheel is a Hoosier "3 x 8." I consider Joe Andrews' HP anvils and upper wheels to be the industry gold standard and I've never met anyone who would argue that.

A LOOK AT THE MARKETPLACE

That old bit of wisdom "Buy cheap, buy dear" was never more true than when it comes to E-wheels. If you want quality you've got to pay for it. If you don't have quality you'll pay for it in hassle, frustration, and failure. These are machines that I recommend considering.

LIGHT DUTY, STANDARD DUTY, SPECIALTY/MULTI-PURPOSE, ANVIL AND WHEEL SETS

High quality wheel sets are available from Justin Baker, Ron Fournier, Metal Ace, Kerry Pinkerton, Hoosier Profile, and Kent White. Due to CNC machining, these companies can now work with you to create custom profiles and custom sizes at surpris-

Main image: Use a 10mm end mill to convert your V-groove into a 13mm deep channel. Inset image: Close up shows the milled channel. Note the pics on the next page to see how we modified the lock bolt for a better fit.

ingly reasonable prices. The key to all anvils and upper wheels is the quality of the bearings. Cheap anvil sets mean cheap bearings and also low grade steel, rather than tool steel. Go ahead and spend a bit more to get your wheels hardened if that option is available. Above all, no matter what anvils you own, don't let them get knocked around and make sure they have a safe place to live when not in use.

IMPROVING YOUR IMPORT

One of the problems often seen on inexpensive e-wheels is yoke slop. There are two ways to deal with this problem. The first is to simply round off the locking bolt as seen in the photo sequence. This eliminates 50% of the free play. The best way is to remove the yoke column completely and re-mill the V-groove into a ½" (13mm) deep channel just a fraction wider than the locking bolt. Since the latter is usually 10mm, you'll want to use a 10mm end mill and make a second pass fractionally to one side. Once this is done, completely grind off the nose of the locking bolt and run it as deeply as possible into your new channel (don't lock it, though). You will have eliminated 95% of the play doing this, and you'll feel you have an entirely new machine. If you don't have access to a milling machine a machine

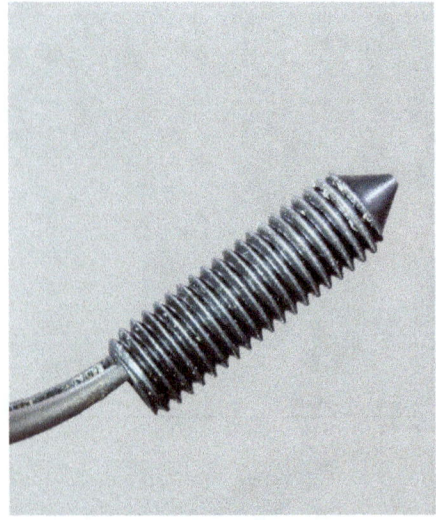

The locking bolt with the pointed end that needs to be rounded.

Chuck the bolt in a drill and spin it against a sanding surface. Use light pressure and swing the bolt side to side.

The result is a nicely rounded bolt that makes better contact in the yoke's V-groove.

shop could do this for you for the cost of ½ hour labor, typically about $40.

Getting the Upper Hand on an Upper Wheel

The machining on some import upper wheels is so bad it appears as if they were cut with dull tools…they probably were. Whatever finish is on the upper wheel will transfer to the panel and so ridges in an upper wheel are disastrous. If you don't have a lathe, or can't go to a machine shop, here is a quick-'n-dirty way to fix the problem - note the picture on this page. Purists will cringe, but it's worked for me. Wray Schelin told me he did something similar once when trying to teach e-wheeling on a junk machine in a poor Caribbean nation. Purists don't know poor.

Lemons into Lemonade

If you've got a good frame there is little need to replace it even if the upper wheel and anvils are low quality. The most economical thing to do is upgrade just the wheel and anvils. It doesn't take much work to convert your machine to use them, even if the upper wheel carrier and anvil yoke that came with your machine are the wrong size.

I stumbled on my Lazze-clone e-wheel at a freight salvage sale and bought it for a song. I used wheels that came with it for a while, but as soon as I could upgraded them to Hoosier Profiles. I have a very nice machine now and 90% of the cost went into the HP upper wheel and anvils. I win!

I built this simple rig to polish a low-quality upper wheel. Get the wheel turning and then work a spinning disc sander across the wheel's surface starting with a 120 grit disc and gradually working towards 600 grit. Don't apply too much pressure. The results are quite good.

Main image: I've inverted the imported machine and cut off the factory upper wheel carrier which had been spaced to carry a 2 in. wide wheel.

Inset image: Using a machinist's block plus a short strip of 1/16th aluminum angle (see it on the right of the block), I jigged up my new carriers for a Hoosier Profiles 3 in. wheel. The 1/16th in. spacer strip is critical so that the new wheel doesn't bind. Tack weld carefully to anticipate warpage before finish welding.

I made these various yoke holders from scrap steel tubing. Only the second from left required any welding, the rest were simply cut. They are interchangeable on my yoke column with two bolts.

Romeo Lalli Builds An E-Wheel

Romeo Lalli had no formal engineering training yet developed this remarkable chain-driven e-wheel in secret to form panels for Sikorsky helicopters. Lower anvil could be raised hydraulically with a foot pedal. The gauge was used to set consistent pressures between wheels during production runs.

Note the anvil resting on the frame. Developing this machine in isolation, and secret, Lalli used outmoded bronze bushings. His innovative "tapered G" frame design presaged Lazze Jansson's by half a century.

IN FROM THE COLD:
ROMEO LALLI'S WHEELING MACHINE

During the Cold War the West depended on its inventiveness to defeat the USSR's overwhelming advantage in numbers of men at arms. One of America's unheralded inventors was Romeo A. Lalli (1906-1987) who set up a machine shop in Stratford, Connecticut in 1935. An untrained engineer, but empirical mechanical genius, Lalli developed top secret machines and processes that were critical to the success of the military helicopters that Igor Sikorsky was building in his nearby Stratford factories.

For the first time, one of Lalli's machines is revealed publicly, a remarkable motorized English wheel with hydraulic anvil adjustment. Powered by a 1 ½ hp electric motor, this machine was built in the 1950s and used to form helicopter fuselage panels, including, it is believed, some from titanium. The lower anvils used bronze bushings instead of ball bearings, but Lalli would have had almost no exposure to the features commonly found on British-built wheeling machines because at that time only one or two were in America on the other side of the country in California hot rod shops. Lalli, because of the nature of his work, had to develop his machine in secret.

A 1 ½ hp motor drove a chain reduction set which drove a geared reduction group attached to upper wheel creating enormous torque. The machine was built in Lalli's shop, which still exists in Stratford, CT

Mitch Bell builds an e-wheel — Build Your Own

You can build your own English wheel if you have a welder. Doing this will save you money if you do a good job of it, otherwise you've wasted time and money. The biggest obstacle to welding your own machine is anticipating how the welds will pull parts to one side as they cool. The simplest forms of English wheels are "C" shaped. It may look easy to just weld a couple of short arms to a taller spine, and presto…you've got an e-wheel. However, a lot of home-made machines end up as scrap because what the builder thought was a carefully jigged frame ended up twisted and splayed. If you do decide to weld your own, start with tack welds and then check alignment. Do some more tack welds and recheck. Deal with twist and misalignment immediately. Flame, or laser, cut steel sides eliminate most weld shrinking issues and are a much better way to go.

If you choose to build your own, I recommend buying the adjustment mechanism. A lot of builders have used inverted trailer jacks, often modifying them a bit to eliminate play in the tubes. I've done this successfully. However, for not much money you can get a well-engineered parts kit and save yourself a lot of time, experimentation, and hassle. Remember: Wheel More, Worry Less.

Mitch Bell, from Scotland is new to e-wheeling and has plans to roll a new body for his Volvo P-1800 completely in aluminum. He and a buddy, Darren Speck, came up with a well-conceived design for a box-section frame which Darren put into AutoCad and then had lasered out of 6mm (1/4 inch) plate. The curves are 3mm (1/8 inch). Mitch welded it together, dealing with distortion as he went along, and now the results speak for themselves, a rugged, hefty e-wheel at a fraction of purchasing a completed one. The adjuster mechanism and wheels are Hoosier Profiles. You can get a laser cut frame kit from Mitch at: abzperformance@gmail.com.

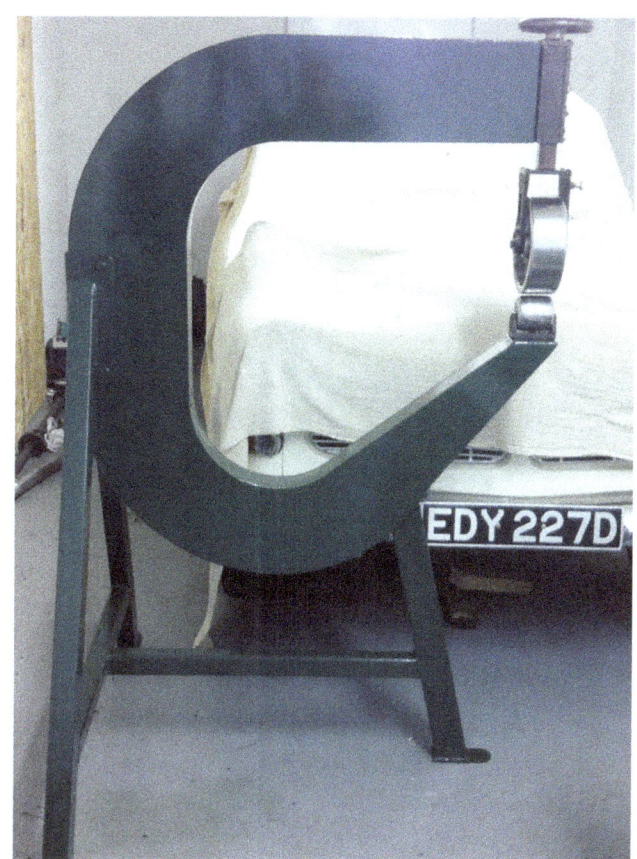

Mitch Bell's home-built machine, using an aftermarket adjuster kit, ready for action. Kit prices are quite reasonable, but the work still requires patience.

A laser cut parts kit - like this one from Mitch Bell - is the surest way to have a successful build. Note the tabs on the internal stiffeners which make for perfect alignment of all parts.

Chapter Three

Getting Started

Get the Basics Right

Before we actually roll panels, let's get the basics right. The following points must be observed for all the rest of the lessons and examples in this book. Failure to remember, and observe, these fundamentals will result in misshapen panels, and fingers.

How to Crush Your Fingers, and How Not To

In the first chapter I said that you could consider an English wheel nothing more than a rotary hammer. It therefore can be assumed that if you get your fingers caught between the upper wheel

A panel should be held at its edges, preferably one hand behind the wheel and one hand in front so that you don't just "shove" a panel through. Ultra-thin gloves from Wray Schelin keep aluminum-oxide off your hands yet allow for excellent feel.

and the anvil, the resulting pain will be like repeatedly hammering your finger. The wound, if not bloody, is usually a deep purple, pulsating, pile of pain… and I mean REAL pain… out of action for three days pain… should I go to the hospital pain… who can I sell this #*!%^@ machine to pain. Most of the professional wheelmen that I have met have flattened thumbs, and scarred adjacent digits. These maimings occurred during their apprentice "party like a rock star" days… usually on a Monday morning and always right after they ignored their mentors' warnings. After all, why should they listen to some old geezer… Yeeahaaaaa!!!!!

Keep your fingers on the edge of a panel and try to hold it only along the outer ¾ inch (18mm). Never hold a panel in a "death's grip" with white knuckles. If you aren't holding a panel gently, it is likely that you'll pull a twist in it as you roll it anyway. If you feel you need to hold a panel tightly because it is heavy and in danger of flopping over the wheel, you should either roll a smaller panel, or get an assistant to support the opposite side. Wheeling is a matter of "feel." You can't properly feel a panel if you're clutching it like it's trying to run away. Since you rarely wheel the edges of a panel, the very edges are the safest place for your fingers.

Wearing gloves is a good idea. One reason to wear gloves is to keep from getting the black, ever-present aluminum oxide that blooms on aluminum panels from embedding itself in your palms and fingers. Gloves also protect you from the danger of slicing your finger along the edge of an unexpectedly sharp edge. Thin Kevlar gloves are the best for avoiding these cuts, but Wray Schelin (see Sources) sells an excellent type of glove which is thin, resists abrasions, and still allows you to feel the panel. I highly recommend getting a few inexpensive pairs of these.

Don't Forget to Wipe

Before inserting a panel between the wheels you will ALWAYS wipe the upper wheel, and the anvil wheel, with a rag soaked in some sort of quick evaporating solvent such as acetone, or lacquer thinner. There isn't a shop in the world that doesn't have dust floating around it and even the

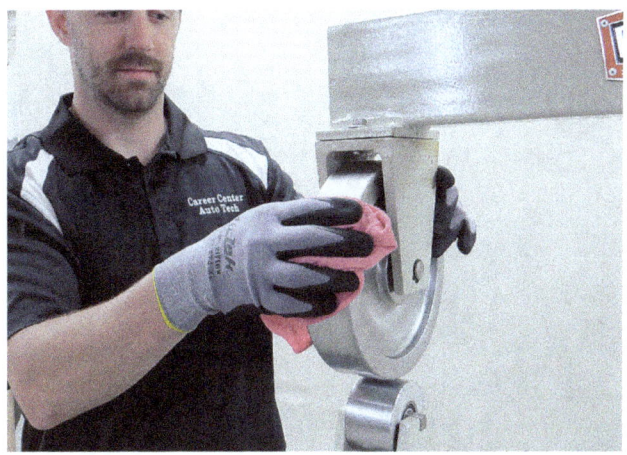

Travis Stewart uses a fast evaporating solvent-moistened rag to wipe the upper wheel and anvils often. Always wipe a panel before wheeling it. Make this a ritual.

If welding steel panels, use a stainless brush to remove any slag before crushing the weld in the e-wheel. If welding aluminum, use a splash of vinegar and a lot of water to remove the residual flux which can scratch your wheels badly.

Always deburr your panels before wheeling to prevent cutting your fingers. Here James McMahon of MPH Motor Panels uses a Vixen body file to deburr, and level, a panel edge.

I prefer a swivel headed deburring tool because they are safe, can cut inside curves, and easily slip in a pocket.

You'll have to pry my B-2 Beverly Shear out of my cold, dead hands! Simply the best cutting tool ever designed.

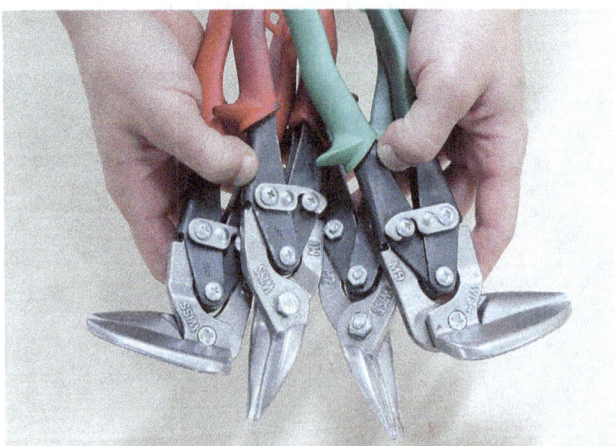

Aviation snips are an industry standard, but the serrated edge they leave must be dealt with prior to wheeling or welding. Wiss, Midwest, and Milwaukee are the better brands.

merest speck can spoil the finish of a wheeled panel. Anything more than a speck of dust, such as a piece of sanding grit, will not only scratch your panels, but will also scratch your vitally important and expensive wheels. DO NOT neglect to make sure, after welding, that there is no flux or loose bits of welding slag on your panel. You should never wheel after gas welding aluminum until you have thoroughly washed, and brushed the weld area and removed the flux. One of the great benefits of gas welding is the easy to crush weld and the ability to quickly re-raise the panel where the welding shrank it by simply wheeling along the weld line. However, if you bugger up your wheels when weld-crushing, you haven't done yourself any good. Don't forget to wipe, and wipe continuously! If you leave the wheel to answer the phone, dust has time to float in and land on your panel. As Devo sang, "Wipe it good!" ...or something like that.

Deburring

Always pull a swivel headed tungsten deburring tool along all the edges of a panel before presenting it to the wheel. I recommend swivel heads because they are the easiest to use on inside curves. Some shops like to use Vixen files (body files) because these can be used to even off a wavy panel edge. If you use one of these be sure to "cup" your hand around the file using your thumb as a sliding guide on top of the panel and your other fingers as a sliding guide on the bottom. Failure to mind all five fingers when deburring a panel this way will result it scrapping over your finger rather than the panel. Because of this danger, I prefer the simple swivel headed tool which is always stored handily in a pocket.

The easiest way to minimize the need for deburring is to cut a panel with a tool that leaves a smooth edge. For long cuts, a stomp shear can't be beat. However, these are pricey tools that take up a lot of room. The greatest cutting tool ever invented is the legendary Beverly Shear, made in Chicago by the same family since the 1920s. They're still making them. It's a rare professional shop indeed that doesn't have one or more of these beauties. You don't know what a clean cut is until you've used a genuine Beverly Shear. Split a line the width of a hair? No problem. The imported knock-offs of these machines are pitiful and not worth the money. They'll have to pry my beloved B-2 out of my cold, dead hands!

In the U.S. most of us cut sheet metal with a good ole reliable set of aviation snips. Av snips have also been around a long time and they are great because they allow very precise cutting, and use compound leverage action to make it easier to get through thicker gauge metal. Their drawback, however, is that they leave a serrated cut line that causes two big problems. Firstly, you have a ragged edge that can cut your fingers as you wheel. It's got to be deburred. Secondly, welding over a serrated edge is bad practice, and can result in a much weakened join line.

The hand snips preferred by English English wheelers are Gilbows. Gilbows come in left, right, and straight versions, just like aviation snips, but these tools leave a cleanly shaved edge. Genuine Gilbows outlast even the best av snips by far, and are easily resharpened. Not only are they a lifetime tool, they are passed down from generation to generation. Knock-off versions exist, but they're for punters. Get the real thing and you'll have fewer issues with burrs.

Layout Guides and Marks

I began working with metal as a member of the Experimental Aircraft Association. That early experience inculcated in me a deep sense of "safety first". One of the rules I leaned was you never write on aluminum with a pencil. Pencil "lead" is actually a very hard material called graphite, and by simply drawing a line on aluminum you thereby create a stress riser which could lead to cracking.

I've been in high end pro shops where it is common to use a pencil on aluminum sheet, but I won't do it. I prefer an ink marker like a Sharpie. I don't use grease pencils although they have been traditionally used, though less so these days. I don't like the residue they leave behind.

Marking a panel, either steel or aluminum, is an ongoing requirement when wheeling. Even the best wheelers need to locate where to raise, where

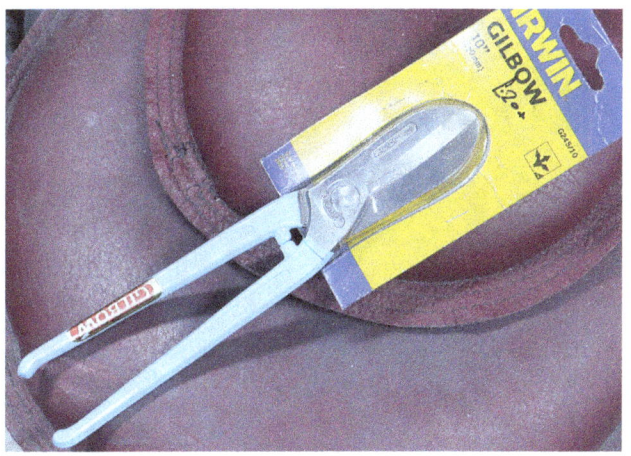

Gilbow hand shears have remained in production since 1885 for good reason. British panel beaters appreciate their velvety smooth cut, and generation-to-generation durability.

Although there are aviation snips in this tool drawer, notice the pride of place given to the six Gilbows. Accept no imitations!

The best method to make notations on a panel is with a Sharpie. They're easy to read and wipe off. Old fashioned grease pencils leave a residue and writing in pencil on aluminum is strictly verboten!

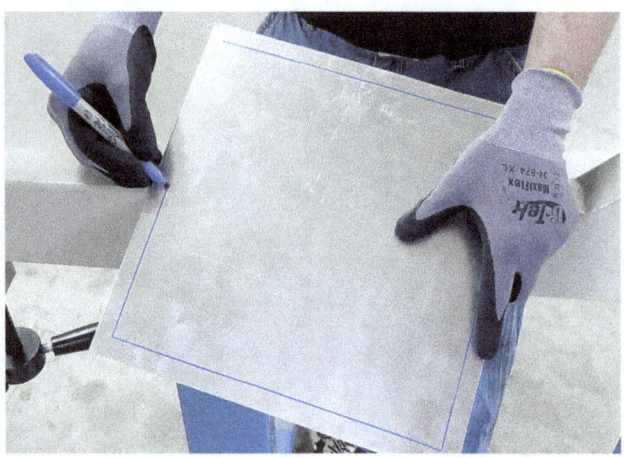

An English wheel is a rotary hammer, the place where it contacts the panel at any moment is called the "blow." Draw a ¾ inch "No blow zone" around your panel before wheeling.

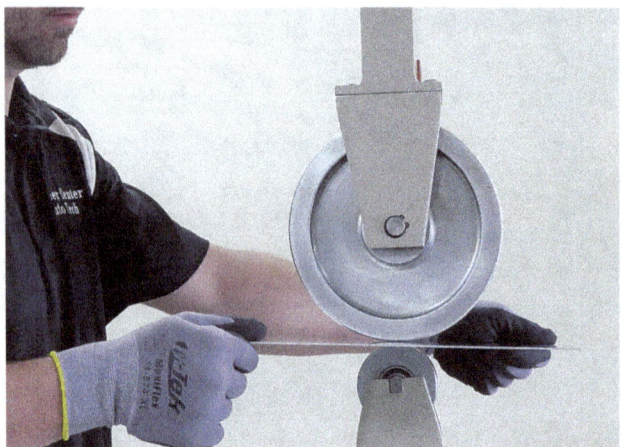

Wheel with the panel at chest level. If your elbows are too high, you'll tire quickly.

Support the panel so it doesn't sag over the bottom wheel. If you wheel with a sagging panel, you roll a twist into it. Because of this, limit panel length to about 3 feet.

to flange, etc. Panel marking has become more of a requirement these days now that our shops are more comfortable. In the old days, especially in Britain and Europe, shops were often so chilly the moisture that accumulated on a panel could be smudged with a finger, and it would leave a sufficient mark to show where work was needed. I think Sharpies are the best marking tool and have developed my own system for using them: Black ink for raising, red for shrinking, and blue for flange and detail lines. If you roll over an ink line, no damage will be done, and the ink will not be pushed into the panel. You can always wipe it off with lacquer thinner. You don't want to weld near ink lines, however. The heat "enamels" the ink making it much harder to remove.

It's good practice to ink a "no blow zone" around a panel. This serves as a reminder where not to wheel. As previously stated, you rarely want to stretch the edge of a panel. An alternative to inking a line is to outline your panel with ¾ inch (18mm) masking tape.

It's All in the Wrist, and Arm, Action

You now know to keep your fingers away from the wheels as much as possible, but you also need to be mindful of your wrists. Don't let your wrists sag down and pull the panel with them. Any tension you impart to a panel while wheeling it will become shape. Any tension imparted when not wheeling becomes a fold. Failure to maintain level hands along a common plane with the lower anvil is the biggest reason for twisted panels.

When I was a teenager I wanted a chopper and so as soon as I got my first motorcycle I threw a set of ape hangers on it. Cool! Until I rode it on long distance trips. Even at that age my arms quickly got tired and I began to one-hand the bars. Dumb and dangerous. I didn't understand about blood flow then, but I do now. You don't want an English wheel where the wheels force you to keep your elbow level with, or above, your heart. You want your elbows a few inches below. This will allow for easier blood flow and you can wheel and wheel and wheel without getting tired. If you have a machine that is too tall for you, stand on a pallet, or some sort of simple platform.

There is nothing cool about a twisted panel which is what you will surely roll if your arms are tired.

Having said that, you should also avoid a situation where the wheels are too low. You'll want the gap between the wheels to be about level with your chest. Find the working height that works for you, but be aware that arms that tire out too quickly are probably due to your elbows being too high.

Team Wheeling

Wheeling is best done as a solo effort with a panel that measures less than 3 feet (1 meter) square. However, there will be some panels, like hood, trunk, and door skins, that require two people to work them. The second person is known as the tail, and that person must clearly understand that under no circumstances is the tail to wag the dog. The lead wheeler does all the decision making, and panel manipulation. The tail simply supports the panel. His or her hands should be on the exact same level as the lead's and create no resistance to the lead's movements. Feet placement of the tail is very important and should mirror that of the lead. The team's footwork must be analogous to a boxer's or fencer's, in that they don't get their feet crossed. If one member falls down, the panel will be creased and ruined.

Geoff Moss told me that his first wheeling job in the early 70s was as a tail in a gang of four at Haslemere Coachworks near London. They rolled large bus roofs with one man at each corner of the panel. There was no joking while a panel was being rolled because the crew was paid piece work. All four depended on each other's competence and attentiveness for their earnings. John Glover, who began two generations earlier, told me the most he ever saw wheel a panel was eight men.

Lead and tail team wheeling. Sean Barton (lead) guides the panel while Jessica Hall (tail) supports it. Don't let your feet get crossed doing this.

Chapter Four

Tracking Patterns

Crowns, Reverses and Transitions

English wheeling is essentially learning how to do three things: 1) roll a crown, 2) roll a reverse, and 3) put a crown and a reverse side by side (transitions). What makes the process an art is that there are an infinite variety of crown shapes, reverse shapes, and "transition" areas.

UNDER PRESSURE

Before you roll a panel experiment with adjusting the pressure your wheels exert on a

1 All three skill requirements are present in this panel: crown, reverse, and transition.

panel. There is no such thing as "plug and play" English wheeling. Rolling a panel is a constant tango between both tracking patterns and pressure adjustment. Each is important. Develop a feel for three levels of pressure.

The first is Light Pressure, where the wheels clamp down on the panel just barely past the pressure where the panel can be moved side to side. In other words, you can only move the panel forward and back between the wheels, but it is very easy to do. This pressure is often called Wash Over pressure and is used to relieve the tensions of a panel that has previously been worked. Forming a panel necessarily creates stresses in the panel which sometimes work against each other, or can be unsprung when the panel is welded, thus causing distortion. A wash over pass after the panel is created, relieves that tension. In the old days when the process for rolling sheet metal stock was not as sophisticated as it is today, panels straight from the mill were deemed to have a "grain" in them analogous to wood. Wheelers would always give a new panel a wash over pass to "break" the grain. That practice is not needed today.

The second level is Medium Pressure. Medium pressure is only slightly more than light pressure but you will be able to feel the resistance a bit more in the wheels. A really nicely adjusted upper wheel and anvil will "sing" under this pressure and within seconds of beginning to wheel you will notice the panel getting shinier and starting to take shape. Slight variations of medium pressure is where you'll work 75% of the time.

High Pressure, the third level, is rarely used because it will put longitudinal furrows in the panel and makes tracking adjustments more difficult. High Pressure is usually reserved for operations such as flanging, or embossing.

2 Light pressure, or "wash over." The panel is just gripped by the wheels and there can be no sideways movement.

3 Medium pressure is the most common setting, and will leave a mirror-like finish on panels if your upper wheel is polished and dust free. Some wheelers chrome plate their upper wheels to enhance their panel's finish.

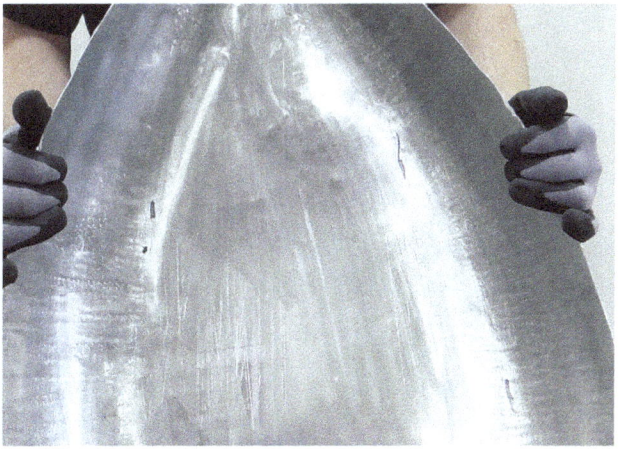

4 High Pressure will leave furrows in panels. It should only be used for specialty applications.

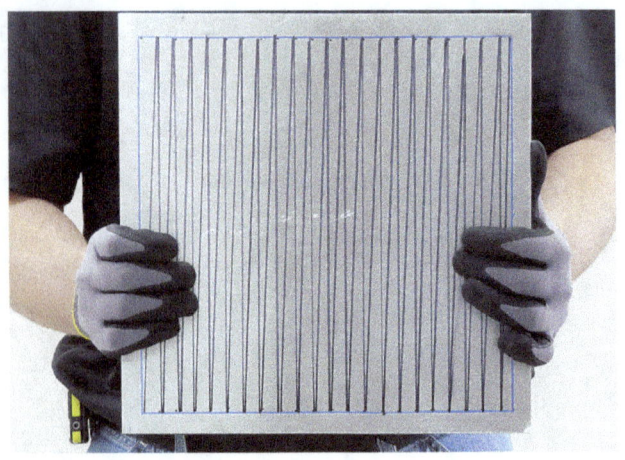

5 The most common tracking pattern is a series of 3/8 inch (10mm) tracks that cross a panel in a series of long, slender "N" shapes.

8 Use the lowest radius anvil you can on a panel. This one is too round.

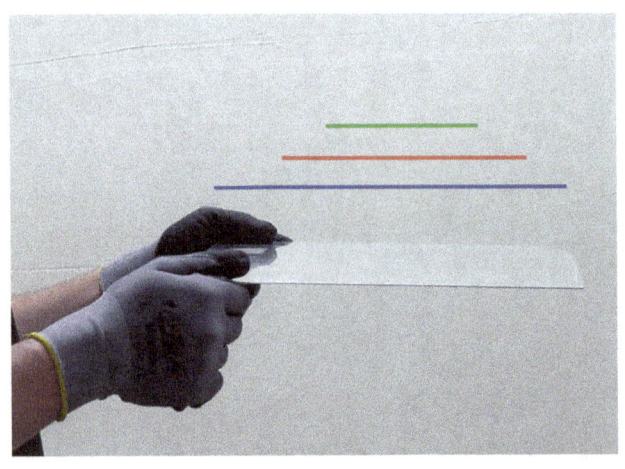

6 The over-lapping Short-Medium-Long tracking pattern will quickly raise a crown in a panel.

9 This one is just right.

7 You can place crown in a panel where you want it by wheeling more in that area.

10 Blending can be accomplished by using pre-planned staggered stops, or a series of random tracks between two topological zones. Here, unstaggered stops caused a ridge which will be blended out in just a few strokes.

14 Another type of cross-wheeling is to wheel across the longitudinal axis of a panel so as to tighten its curvature without stretching it perpendicular to the axis. This common form of wheeling requires a light pressure. If you close-track during this, you can achieve a beautiful mirror-like finish.

Basic Tracking Patterns

In the simplest type of tracking pattern, the wheeler makes a series of long, slender "N" tracks across a panel from left to right. The tops and bottoms of these tracks are spaced, ideally 3/8 inches (10mm) apart. Doing this on a panel with medium pressure will stiffen the panel, and create a small crown.

Now, turn the panel 90 degrees and do the same thing. This is called Cross Wheeling. You'll get an even more pronounced crown.

If, however, your tracking is uneven, or you vary the pressure as you go along, you'll wind up with a wavy panel. Your panel edges will be very wavy if you accidentally roll out onto them at this stage. You should keep away from the ¾ inch (18mm) "no blow zone" you've marked on the panel. The term "blow" in this case refers to the area the lower anvil tracks

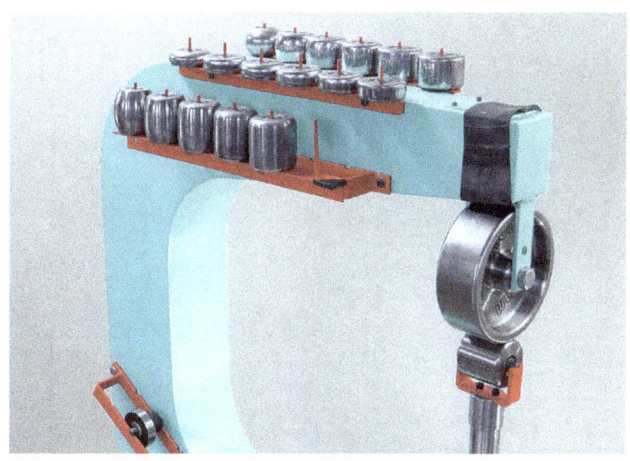

11 A variety of widths of anvils allows you to get into tighter areas. I use 3 inch, 2 inch, and 1 inch sets.

12 I also have a variety of specialty anvils to create specific shapes.

13 One type of cross-wheeling is to turn a wheeled panel 90 degrees and wheel it again.

15 High Crown: Sean starts with a square panel.

18 Then returns R to L maintaining close tracking.

16 He uses the No.2 anvil which is just off flat.

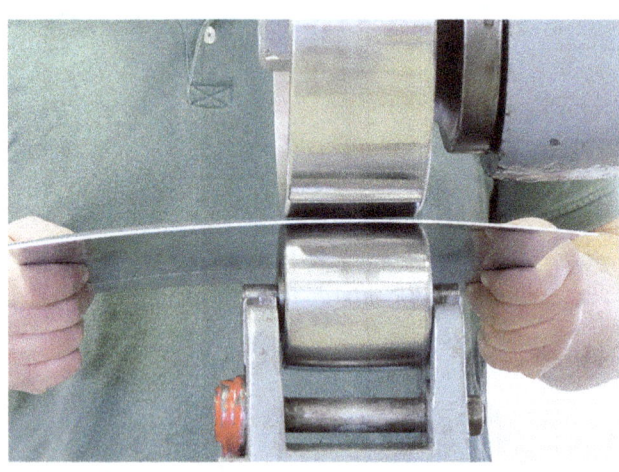

19 He releases the anvil pressure, rotates 90 degrees, returns to medium pressure and makes L to R, R to L passes.

17 Using medium pressure he tracks across it L to R.

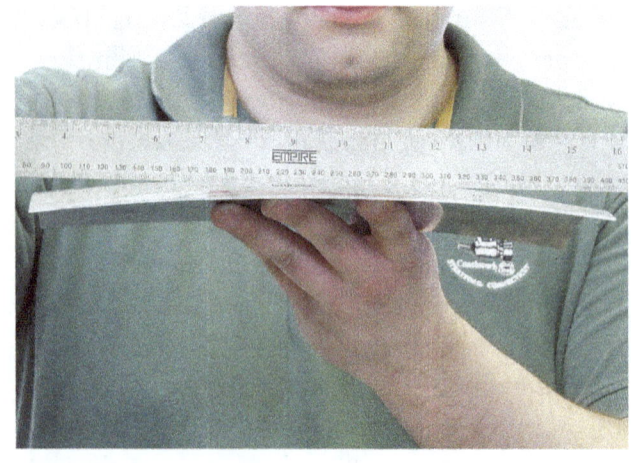

20 The panel has a low crown.

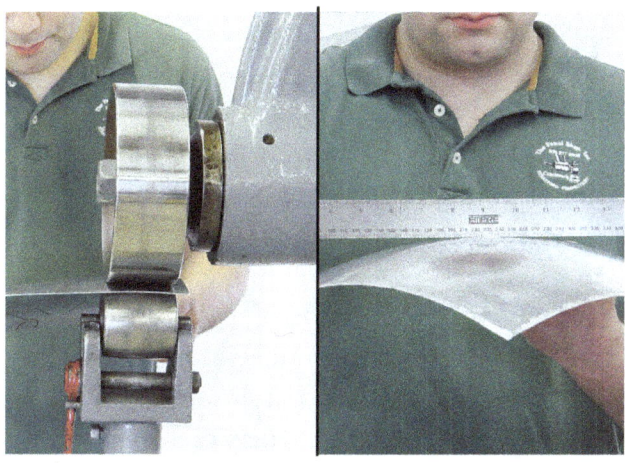

21&22 Repeating the passes he's just done, but using stacked passes each time (short, med. long). RIGHT: Note how quickly the crown develops. Close, evenly spaced tracks gets you a symmetrical, polished, panel.

23 Disc Shapes: Start with a disc…and a smile.

24 Jessica uses a compass to scribe an inner line (I'm going to make believe I don't see her using a pencil!)

25 She begins to wheel staying between the line she drew and ¼ inch from the edge. She doesn't want to stretch the edge or the panel with flop.

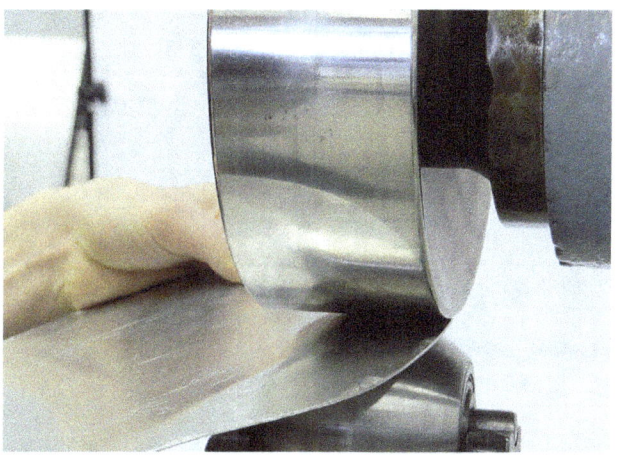

26 Using a back and forth motion, going about 20 degrees of the circle at a time, she begins to create shape as indicated by waviness. NOTE: Though it looks like she's at the edge of the panel, she's not.

27 It would be so easy to crush fingers doing this, but Jessica has been around wheeling machines her whole life and knows where to position them.

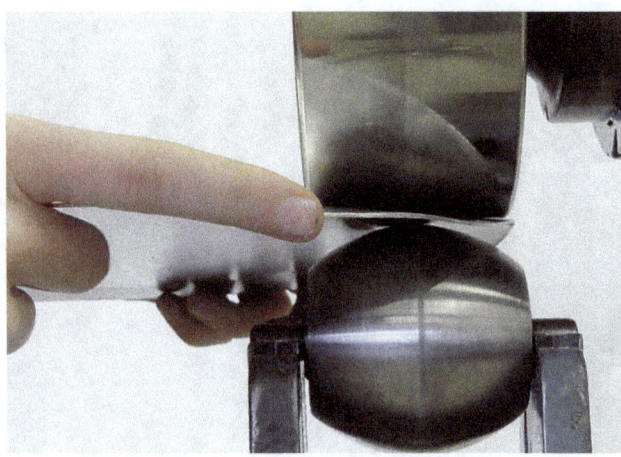

28 Notice the narrow "blow" of even this flat-cut anvil.

29 The waviness in the panel indicates that the edge is stretched enough, but inboard of it needs to be stretched more so as to bring the edge over.

30 Time is money to a professional panel beater, so some quick shrinks on the Eckold make "cents".

upon. The term derives from the "hammer blow" because, as we've said already, the English wheel is nothing but a rotary hammer.

Once you've practiced making consistent 3/8 inch (10mm) tracks across the panel from left to right do the same thing from right to left.

Once you begin to develop a muscle memory for this basic pattern move on to the second most common tracking pattern, overlapping tracking. Here, you will wheel each "leg" of the "N" shape three times before sliding over to the next leg. Each leg will get a "short, medium, and long" pass. This pattern sounds very easy to do, but really is quite confusing unless you practice it a lot. Don't get frustrated. The point of this pattern is to quickly raise the center of a crown. Obviously what you're doing is going over the center of the crown three times and thus raising it more than the surrounding area. If this pattern drives you crazy, there's a "cheaters" way of doing it. Go L to R over the panel doing just long tracks. Come back to the starting point and go L to R doing just medium tracks. Finally come back to the start and go L to R doing short tracks. This approach takes longer, but it works just as well. Be careful to stay out of your "no blow zone" and release the anvil pressure before moving the panel back to the beginning. Set your pressure the same for each pass.

Remember, wherever you wheel, you will create shape. Avoid wheeling where you don't want shape, and wheel more where you do. USE A CONSISTENT PRESSURE during your work so that you come to know what the results will be. You will always get better results by using the LOWEST pressure that will get the job done. Combining the lowest pressure with a close tracking pattern, as well as using the lowest crowned anvil, gives you the best panels.

31 Notice how the edge has tightened right up.

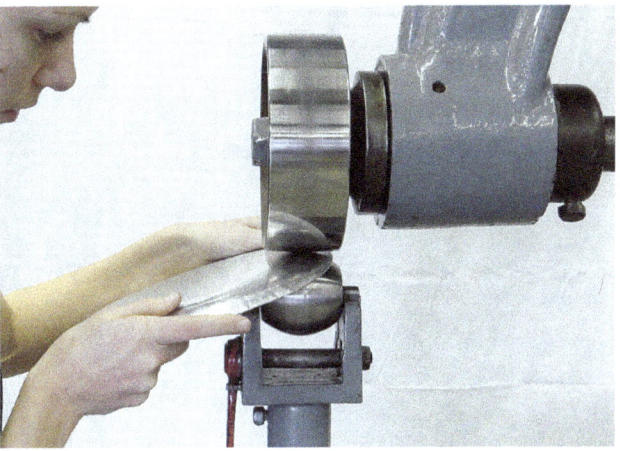

32 Back to the wheel using light pressure to blend the ridge created by the Eckold.

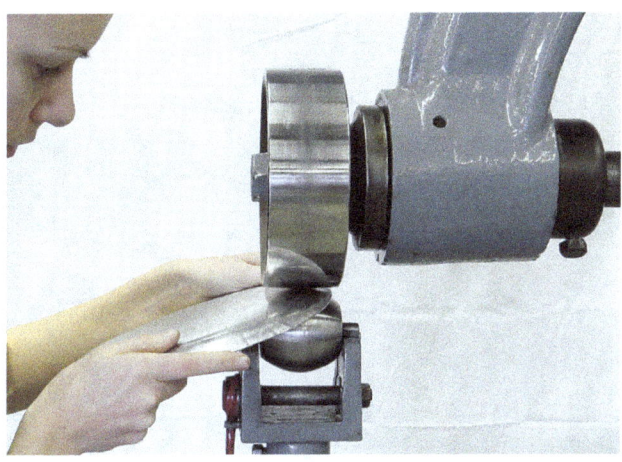

33 The narrow flat on this anvil allows Jessica to move the panel in and out for a better blend.

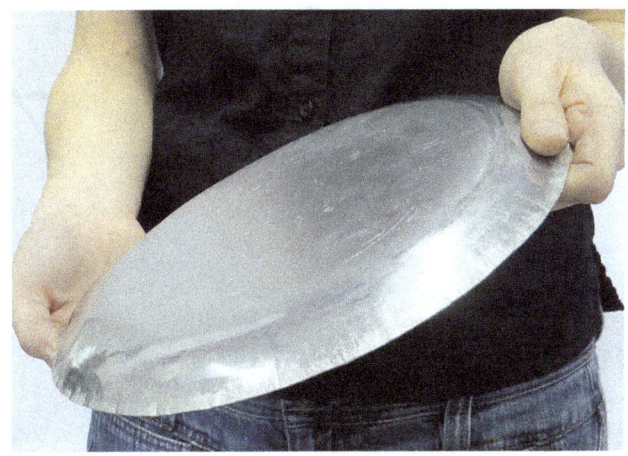

34 This well-shaped panel requires a minimum of sanding to remove the remaining shrinker marks. It is ready to be welded.

BLENDING: TWO METHODS

Once you understand how to set the pressure, choose the proper anvil, and make consistent tracking patterns, you can easily crown any consistent shape. However, what happens when you need to put inconsistent shapes in a panel? For instance, a highly crowned area next to a less crowned area? The answer is Blending. If you did not blend, the panel would look lumpy or faceted. An example of this is when you did your "short/medium/long" panel. If you stopped with absolute precision at each line as you went across the panel, you would have something that has linear ridges across it. However, it is most likely that your stopping points were inconsistent, and so you accidentally blended the panel. By varying the stopping point of wheeling passes you are creating Staggered Stops. One method of blending is to wheel into staggered points between two areas on the panel to lessen the differences topologically.

35 Rolled Edges and Avoiding Hollows: Sean starts wheeling side to side using a No.3 wheel.

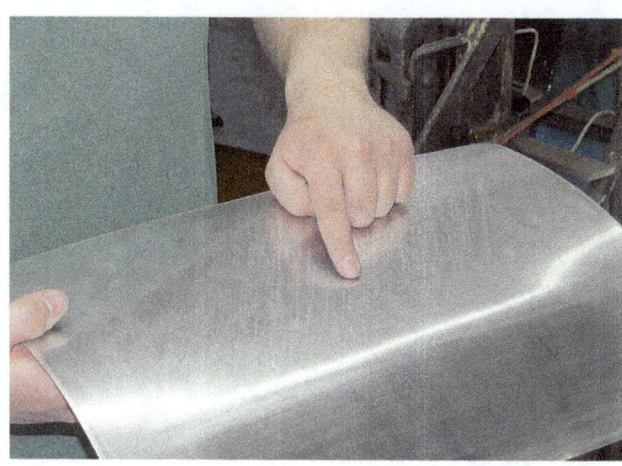

38 …he winds up with a hollow.

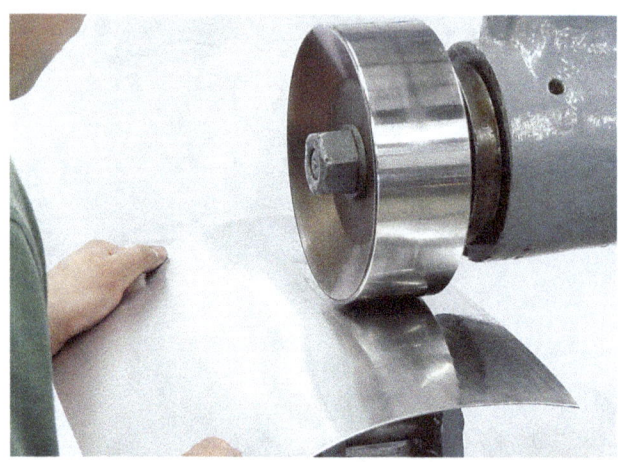

36 He applies some downward pressure to induce the curve faster…

39 The best way to get rid of this is to change to a flat anvil and wheel the hollow up.

37 …but as the curve forms…

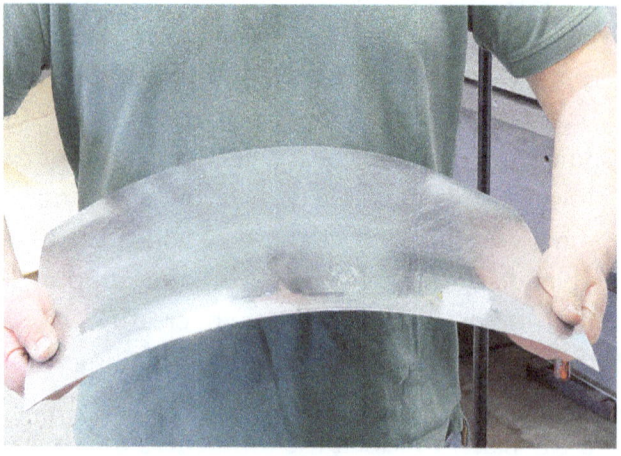

40 After just a few passes the hollow has been raised yet the crown is still there.

Another method is to vary the tracking width. I've described the basic 3/8inch (10mm) track for raising a panel, yet by making tighter, or usually wider, tracking path, you can soften transition areas and create a blend. To change the profile of a crown simply widen the track as you go further from the crown's peak. Widening the track gives you a more gentle slope.

Blending is a routine procedure in wheeling and is accomplished by eye and good judgment, not by any set rule or tracking pattern. Be careful not to over-blend. Sometimes as few as one or two passes into an adjoining area is all it takes.

Anvil Choice

Always choose the anvil whose radius comes closest to the panel's you're forming. You will be surprised how little radius is needed to form most panels. Don't be fooled into thinking a tighter radiused anvil will cut your workload and raise the panel faster. What will happen is you'll put too much shape in the panel before you know it, and then you'll have to face the very difficult prospect of shrinking the panel. Go slowly and raise the panel to where you need it.

Cross Wheeling Again

Sometimes you'll need to tighten up the profile of a panel without imparting much shape in it. This can be done by another form of cross wheeling. In this instance, you use the flattest anvil possible, often the No.1. The danger in cross wheeling come from the tendency to overdo it and not just tighten the panel on the front and back side of the anvil, but also to curve it more from side to side. It is important to do this pass with nothing more than a "wash over" pressure. If done well, you'll have a beautiful mirror-like finish on your panel. Real pros anticipate a slight pulling in of the panel fore and aft, and side to side, so they rough wheel their panel to about 97% before doing this pass. They let this pass bring in to the final shape they're after.

41 Reverse: Start with a flat 12 inch x 18 inch (30cm x 45cm) sheet.

42 Jessica uses slip rolls here to put a curve in the panel quickly.

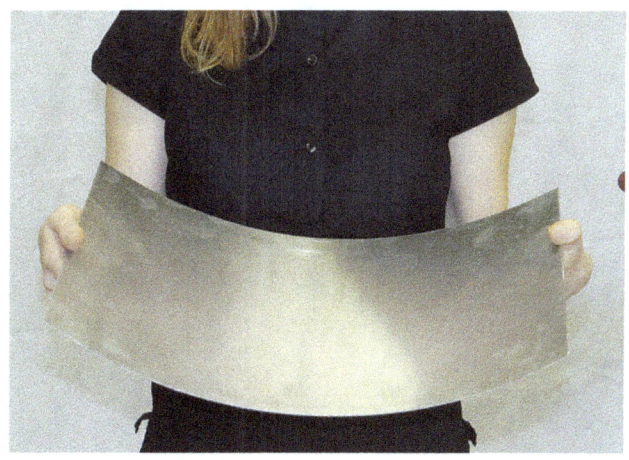

43 The panel now has a curve going longitudinally.

44 She changes to the No. 4 anvil (medium radius).

47 She makes more passes at the edge of the panel and diminishes them as she comes inboard.

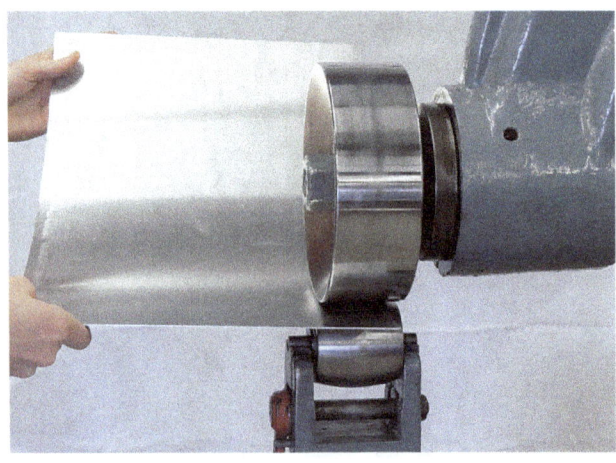

45 The panel goes in about 2 inches (52mm).

48 Here Jessica lessens the pressure as she comes inboard. One of the advantages of a foot wheel is you don't have to take your hands off the panel.

46 As she begins to wheel it, she pulls down on the panel causing the flange to curl away from the top. Pull evenly with both hands and don't jerk it.

49 After only a few minutes she's got a nicely formed reverse going.

Note: Photos 1-14 are low crown. 15-20 are high crown, and photos 21-22 show disc shapes.

DISC SHAPES

Raising a "hub cap" dome in a disc requires the same basic operation as raising a crown in a square. Stacking your passes over the center will raise it more quickly and give you a more pronounced center. With a disc, you simply rotate it a few degrees after each pass and thus work your way around the circle. Light pressure and more passes will give you a more consistent dome.

However, disc, or semi-disc, shapes are often required next to flat areas. An example would be making an inner wheel tub, or even a headlight bucket mounting surround on the front of a fender. Here's how to do it.

ROLLED EDGES AND AVOIDING HOLLOWS

Photos 23-34. There are many panels on a car body that require a rolled and crowned panel immediately next to a fairly flat panel. For instance, the front fenders on many cars from the 1950s. These had high crown tops leading to the headlight, but on their sides they're slab-sided. Rolling them on an English wheel in one piece will cause you to pull a large "hollow" on the slab side as the crown develops. Here's how to deal with that.

REVERSE

Photos 35-40. Reverse are the second most common shape in panel forming. They are literally the "reverse" of a crown. A crown can be turned over and it becomes a "bowl" which holds water." No matter how you hold a reverse, it cannot hold water. The process to create a reverse (sometimes called a Saddle) can most easily be stated as simply stretching the edge of a panel. Stretch less inboard of the edge, and more the closer you get to the edge.

MATING CROWNS AND REVERSES

Photos 41-52. A second way to roll an edge is to simply shrink it using a mechanical shrinker. This is very fast, but will still pull a hollow in your panel that needs to be raised. Once you do that, wheel the edge to get rid of

50 At his point, her elbows are out and she rolls her shoulders to move the panel in a circular sweeping motion.

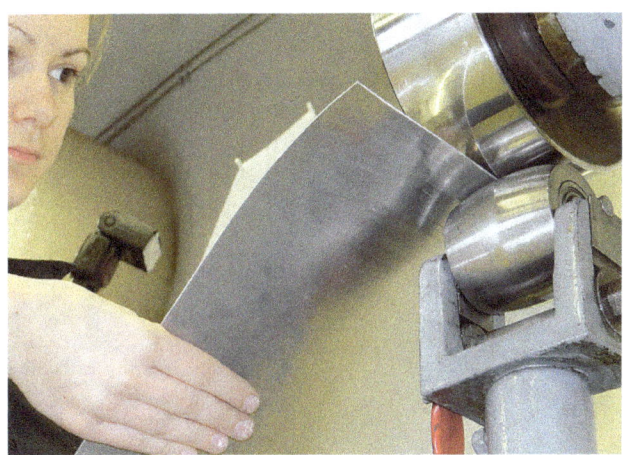

51 The depth of the reverse is limited by the width of the anvil.

52 The resulting polished reverse is due to patience and close tracking.

Mating Crowns & Reverses: Mechanically shrinking edge of a panel is fast. Eckold Handformer is the prince of hand-operated shrinkers. RIGHT: Stippled jaws don't tear the sheet as serrated jaws may, so you are free to wheel again. You can get your machine's jaws stippled by Neil Dunder (see Sources).

57 …and after a little Mk 1 knee manipulation…

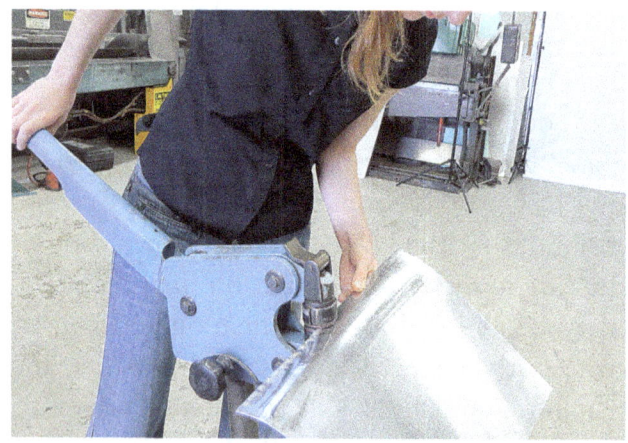

55 A little wheeling will bring up the hollow and smooth the edge's curve.

58 …and a little shrinker help from Jessica…

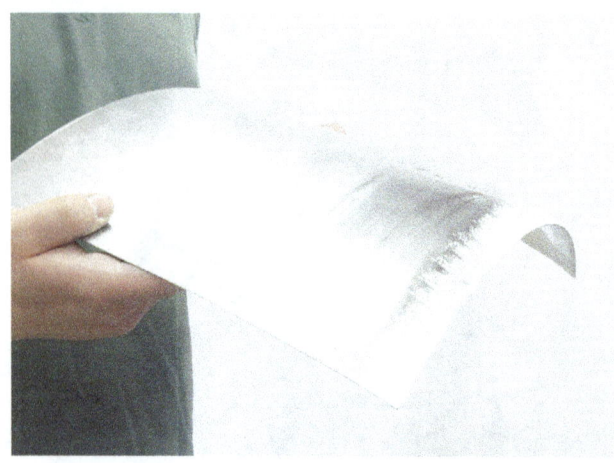

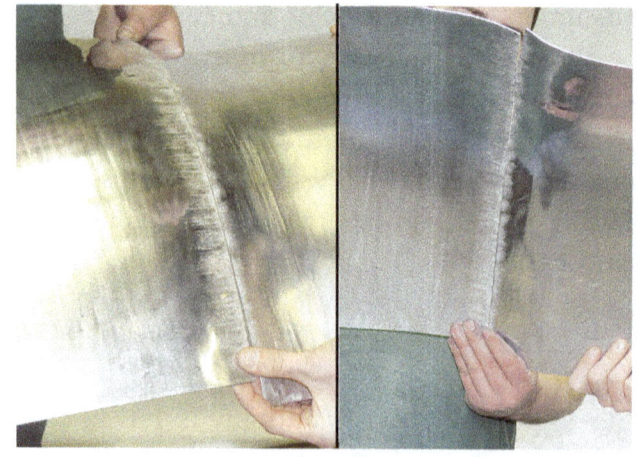

56 And now, just for fun…

59&60 …we have two panels that fit perfectly. RIGHT: Mating crown and reverse panels are the fundamental requirement for most body panel work.

61 *Anticlastic Reverses: Sean uses two "C" clamps to fix a 1 ½ inch (40mm) pipe to the end of a worktable.*

65 *Sean begins to stretch one of the flanges, few passes inboard, more outboard.*

62 & 63 *He and Jessica now press an aluminum blank over the pipe using even pressure. RIGHT: The result is a trough with a gentle radiused valley. How easy was that?*

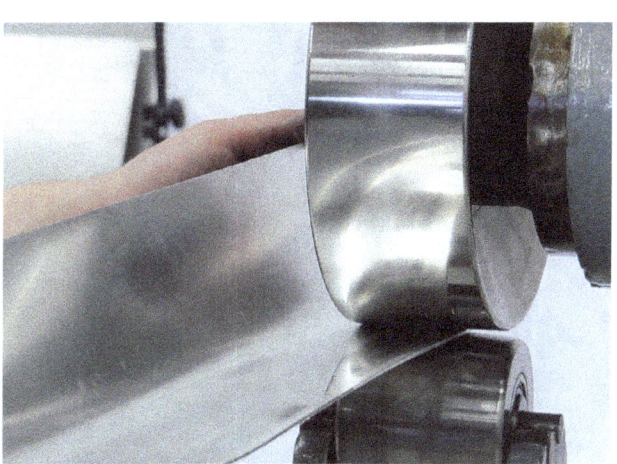

66 *After a few passes already the flange edge is getting wavy (lower center left).*

64 *The No. 2 anvil is inserted in the yoke, note thin strip of aluminum under L side of axle. Cocking wheel this way allows for easily graduated pressure - flange of blank will not go much past the center of the anvil, and will be squeezed more along its edge.*

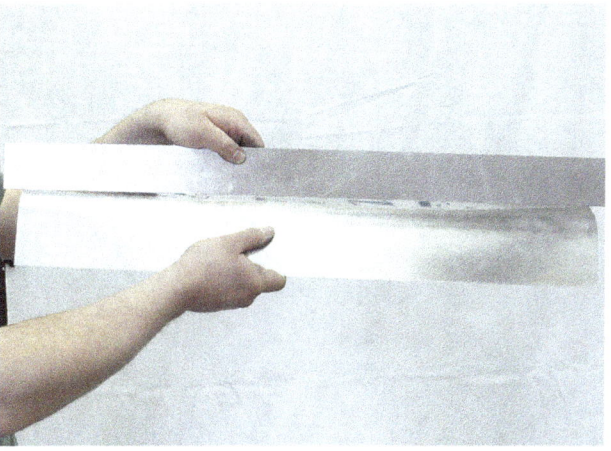

67 *A straight edge shows the anticlastic curve forming in the valley.*

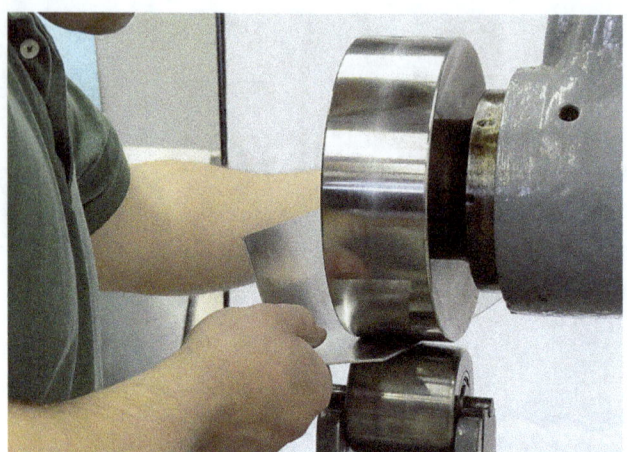

68 The process is repeated on the other flange.

69 More wheeling of the flanges creates an even more pronounced reverse.

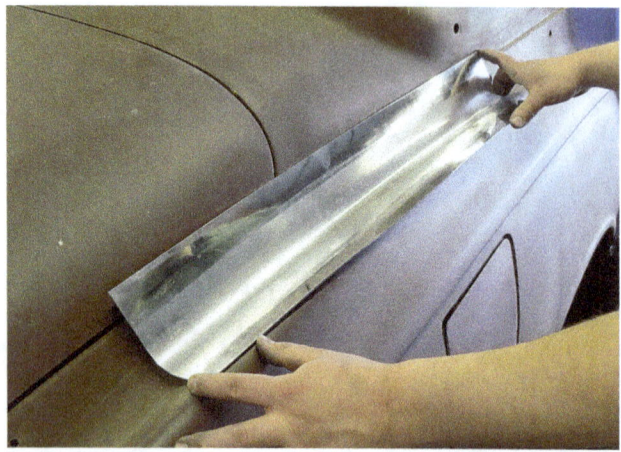

70 In less than five minutes the double reverse is formed. Sean holds the panel against a Rolls-Royce Phantom V quarter panel.

the ridges left by the shrinker and give the panel's edge a consistent roll.

ANTICLASTIC REVERSES

Photos 53-60. Another fundamental requirement for autobody construction is the anticlastic panel, a panel in which the curves go away from a common point. Crowns, by the way, are technically called synclastic because their curves go towards each other.

Almost any time a fender or quarter panel joins the main body of car you'll find this type of panel. They are also very common around grilles, headlights, and the cowl where the windshield joins the hood. Though they look extremely difficult to make, they are nothing but adjoining reverses of the type we demonstrated earlier. You can wheel them, or stretch them in a stretcher and use an English wheel to smooth them.

TUNNELS

Photos 61-70. If you don't have a set of slip rollers, you can use your English wheel to create simple curves in your panels (versus compound curves). One method uses the standard steel upper wheel, and the flattest steel roller, the other uses a go-kart tire, or similar inflatable wheel in place of your upper wheel and a highly radiused lower wheel, not necessarily an anvil.

SLICK TRICK: The e-wheel can also act as a slip roller to form simple curves and folds (tunnels) in panels. By replacing the steel upper wheel with a go-kart slick, or other soft wheel, a panel will quickly assume the shape of the lower anvil as it is pressed into the pliable upper wheel. It is also very easy to create unique profiles using this set up. Just turn custom lower anvils on a wood lathe out of plastic, or a hardwood like oak. Doing this, you can turn your e-wheel into a low-buck beading machine which can form anything from floor pans to custom side-molding trim. When making tunnels use low pressure, but when beading crank the pressure up. Wheel only in one direction, not back and forth as in normal wheeling.

71 Tunnels: Using the No.1 (flat) anvil and very loose pressure, she inserts a panel about halfway in between the wheels and pulls down and out.

74 A good tunnel is formed whose radius could be tightened, or eased, by more wheeling.

72 She repeats this downward dragging motion moving along about an inch (26mm) at a time.

75 Jim Russell and Rob Banman use Randy Ferguson's e-wheel with a soft upper wheel and plastic lower wheel to curve a panel for Mark Stuart's Daytona Cobra.

73 The panel develops a consistent curve due to close tracking.

76 Ben van Berlow uses a small inflatable wheel to quickly rough shape panels. Others use go-kart slicks. Keeping the tire pressure very low works best.

45

Chapter Five

An Aston-Martin Apprentice Panel

Shape In - Shape Out

From the Middle Ages in Britain up until the 1960s formal apprentice programs were the institution by which boys, mostly, were taught a skilled trade. Apprenticeships usually began at age 14 when the boy's father signed letters of indenture binding his son to a tradesman for typically seven years. Apprentices started out by doing menial tasks, but as they observed their masters, and developed skills, they were given more advanced work to do. The faster he learned, the faster he moved up and the better he was treated by his master.

Before you roll, cut the panel squarely as Amber Burchette does here. Amber races a Limited Late Model when she's not in the shop. #2 .INSERT: Deburr edges using a Vixen file, 50 grit sanding disk, or a tungsten deburring tool (my preference).

Geoff Moss told me that when he was an apprentice at Aston-Martin his mentor gave him a simple task to perform each morning called the "Shape In-Shape Out" exercise. He was required to take a 12 inch (312mm) square panel of aluminum or steel, and wheel an even crown into it, and then wheel the panel back to virtually flat with no "oil-canning." Sound impossible you say?

Wheeling a panel involves thousands of instantaneous decisions regarding pressure, position, tracking, and anticipation. To wheel you must be able to correct inevitable mistakes. Flattening a wavy or lumpy area is a routine skill that is best learned through the Shape In-Shape Out exercise. 80% of everything you need to know about wheeling will be learned in this exercise, so do this and take it seriously.

Can you get a 100% perfectly flat panel? Yes, if you are willing to spend hours. One of the lessons this exercise was meant to teach apprentices is that, "Perfection is the enemy of the good enough." Your goal should be to create a 99% perfect panel. In the real world of "paid by the piece" panel beating, wheelmen couldn't afford to go for perfection on any panel. I've heard this trade secret from many panel beaters. Their customers wouldn't notice the difference and wouldn't be willing to pay the money to achieve it. So keep it real.

Speaking of real, is the panel now back in its original shape and size? No. In reality the panel will be about 2% larger in area than it started out.

Drag a soft cloth along the edges to confirm there are no snags. RIGHT: Wipe the panel to remove dust and grit. If steel, it will be oil coated so use a solvent moistened cloth.

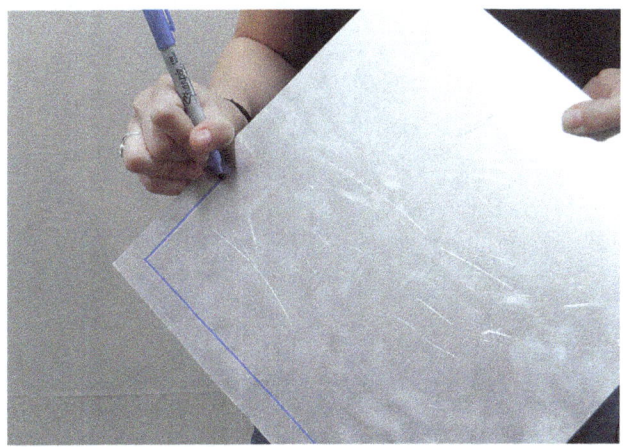

Mark a ¾ inch (18mm) "No blow zone."

Don't forget to wipe…both wheels.

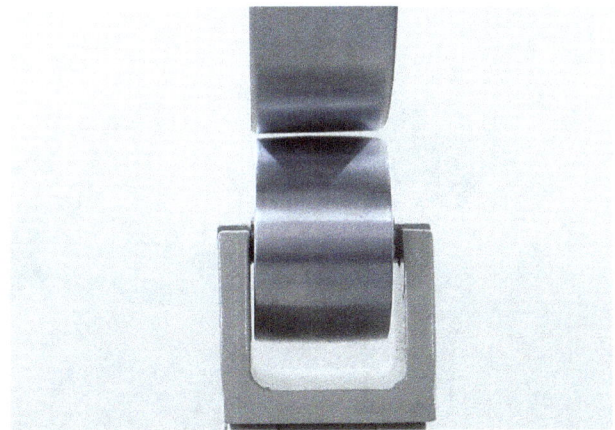

Check that your No. 2 anvil is square to the upper wheel. Paper shims under one side of the axle are commons fixes for many machines. This Baileigh EW-28 is dead on.

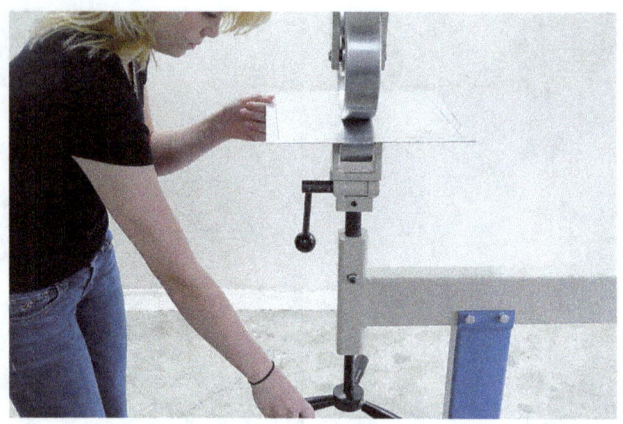

Place panel between the wheels and bring anvil up so it just kisses the panel, you are able to slide the panel sideways. By adjusting the anvil up a bit so the panel is gripped and cannot be slid sideways, you've have "wash over," or low, pressure.

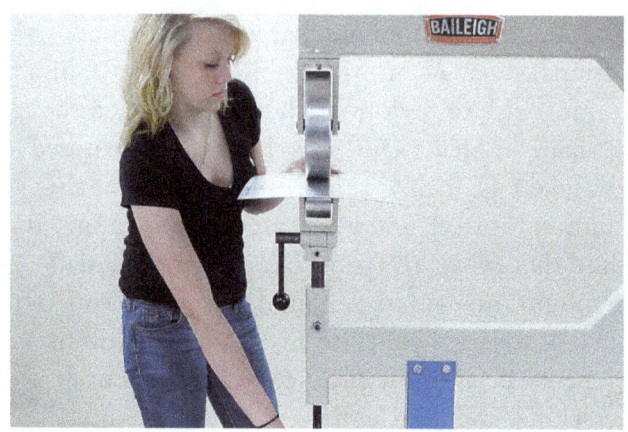

Reinsert the panel and move up to medium pressure, and then repeat the first tracking sequence.

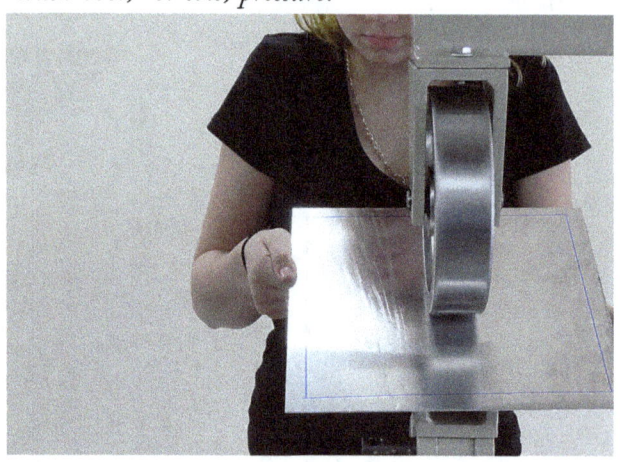

Track left to right with consistent 3/8 inch (10mm) spacing between tracks. Developing a "muscle memory" for this is harder than you think.

Note a crown beginning to develop. If you used more pressure, crown would be more pronounced - with ugly furrows top and bottom. Better to make more passes with light pressure, than fewer with heavy pressure.

Lower the anvil and take the panel out. The panel will be slightly stiffer than before and have a slight front to back curve, but there will be no left to right shape to it.

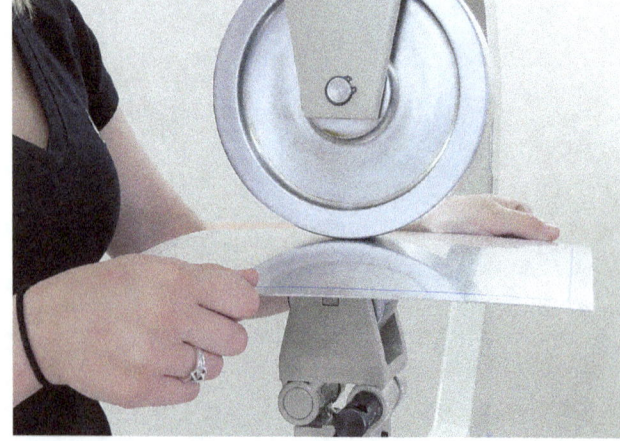

Your hands should follow the arc of the anvil, both front to back and left to right. Don't let your hands "weigh down" on the panel or you'll induce a twist in it.

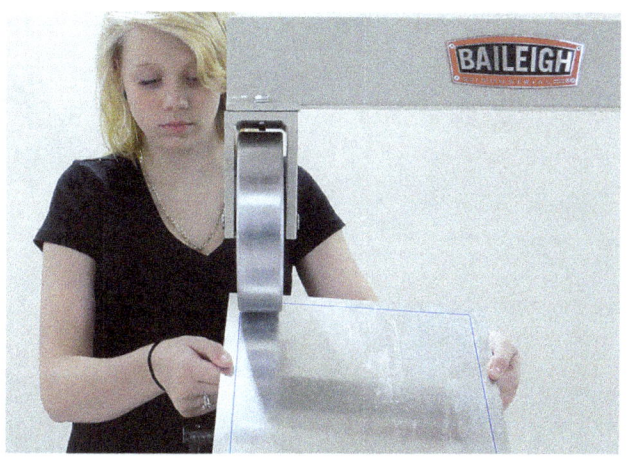

Back to medium pressure. Track right to left this time…

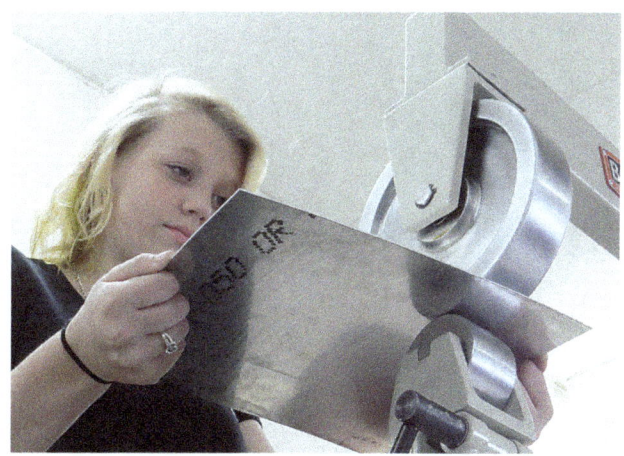

You can "spot" fix your low areas by wheeling them more (light pressure).

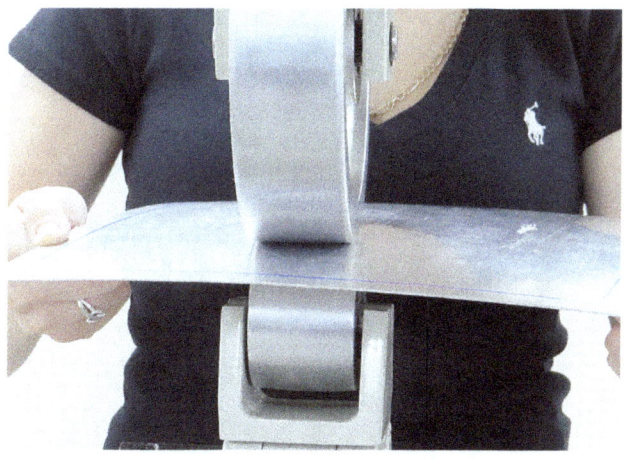

…and then release the anvil pressure, turn the panel 90 degrees, go back to medium pressure, and track again left to right, and then right to left.

To bring the panel to a higher crown, you should re-wheel the whole panel again, but this time use a Short/Medium/Long pattern though again with 3/8 inch (10mm) tracks.

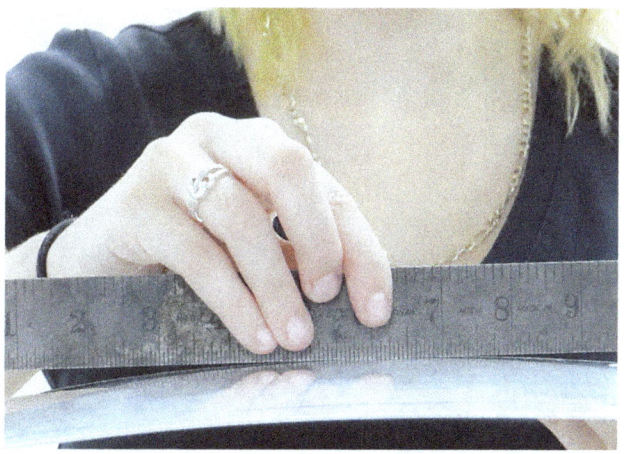

The result this time should be an evenly low crowned panel. If your crown is off-center, it is because your tracking pattern was uneven or you let your hands sag.

Your panel will now have a high crown.

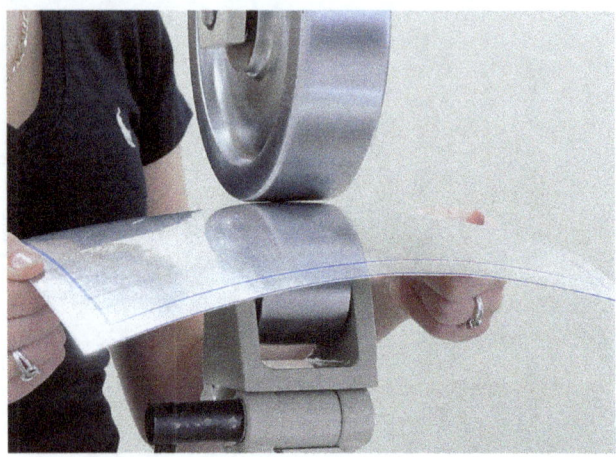

Lower the pressure to "wash over" and wheel the entire panel left to right, and then 90 degrees left to right. This will even out the tension.

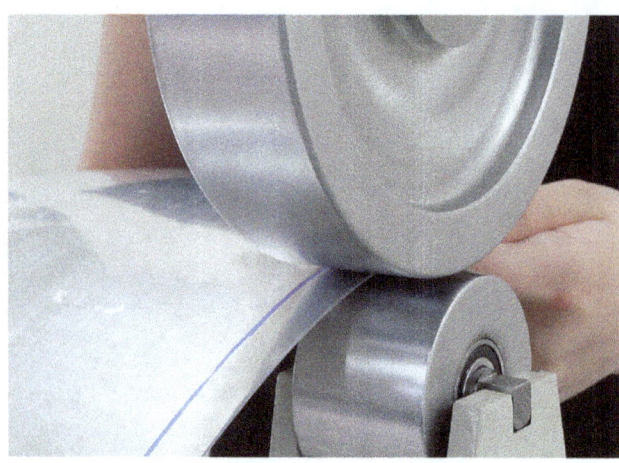

Begin by wheeling in ½ inch (13mm) inboard on all four sides, yet avoid the corners.

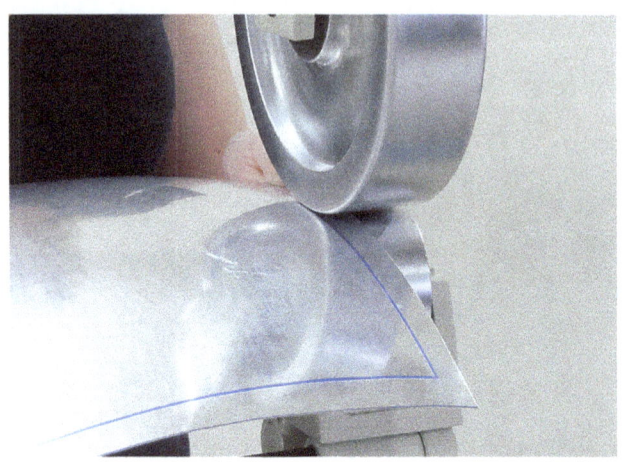

To flatten the panel you have to stretch the sides. The further from the crown you are, the more stretching is required.

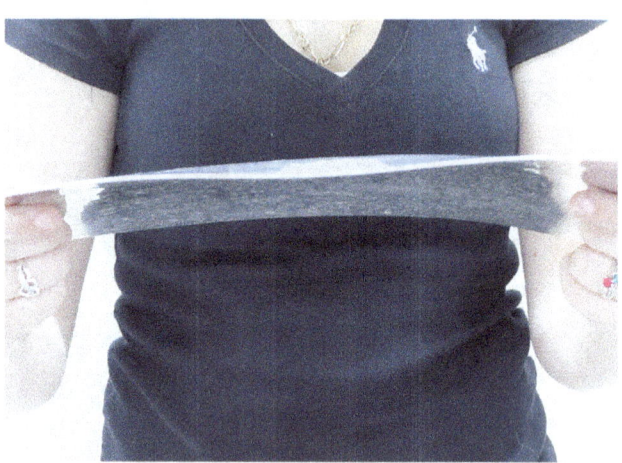

Your wheeling will create a slight waviness to each panel edge. That's good!

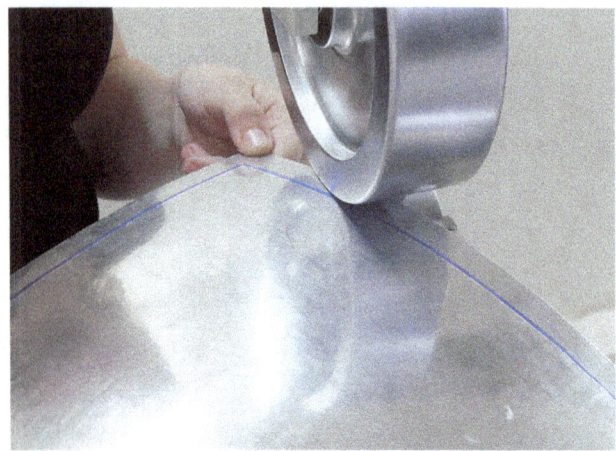

Avoid wheeling the corners, or you will over-stretch them.

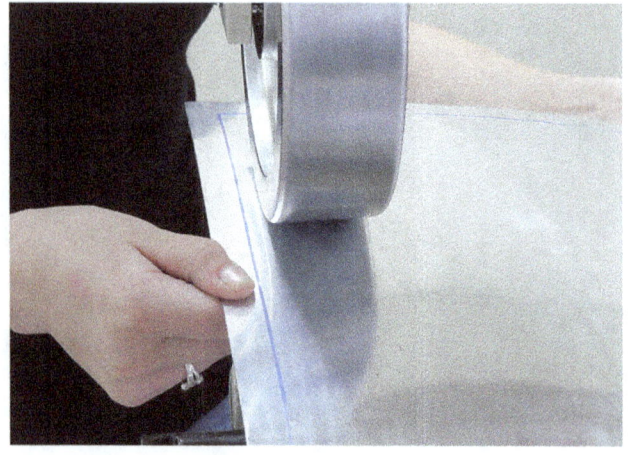

Now move inboard and wheel a 2 inch wide "box" around the panel's edges. Stay ¼ inch away from the edges and again, don't hit the corners.

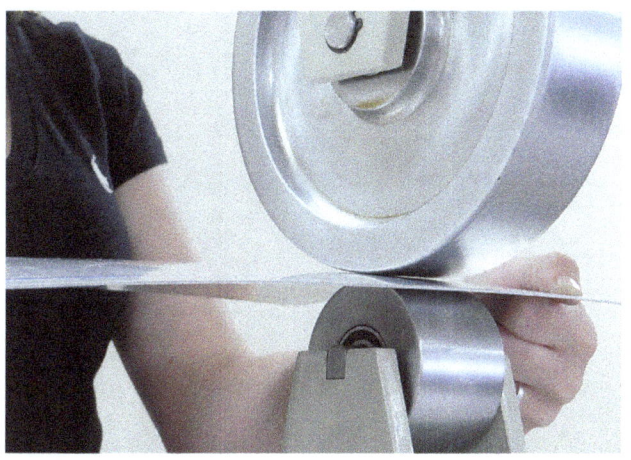

This wheeling is actually creating panel area by squeezing out the metal here. You should notice the waviness of the edges disappearing, as well as the crown in the center of the panel flattening out.

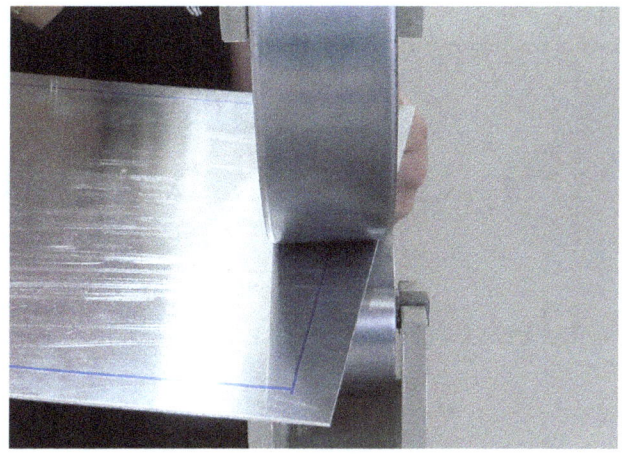

Come at a tight area by angling the panel as you push it into the wheel. It is remarkable how little pressure and movement it takes to give the panel a bit of a stretch and end the oil-canning.

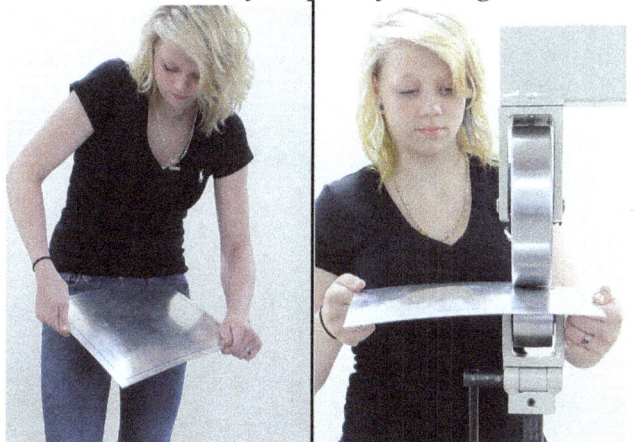

If the panel develops a twist during your progress, lower anvil, remove panel, and hand manipulate it back into shape. RIGHT: Repeat 23-27 until panel is near flat. You can get better results by slightly lifting the panel during the process.

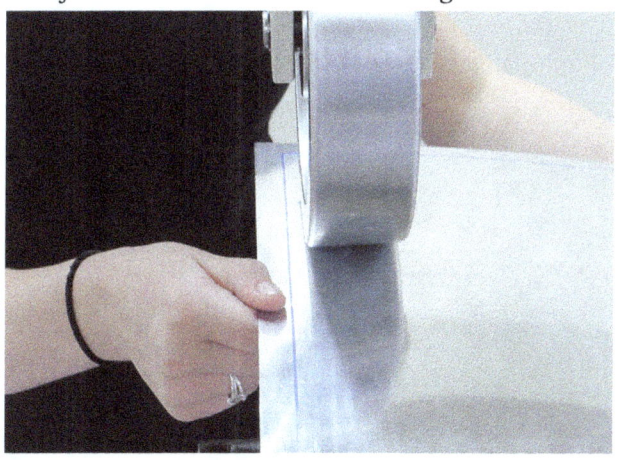

Once you get your panel at, or near, flat, give it a wash over pass to even out the tension.

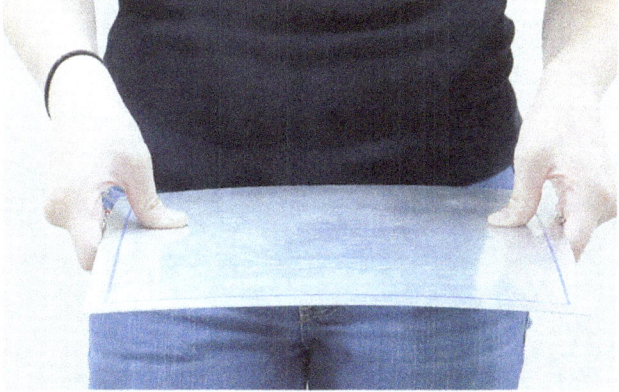

Check panel for any "oil-canning" by flexing opposite corners. Sudden pops mean that you have a "tight" area next to a "loose" area. Loosen tight area by giving it a swipe or two with light pressure on wheel.

DONE! INSET: Amber is a high school honor roll student whose career as a champion kart driver earned her a seat in Limited Late Model racing.

Chapter Six
Building Your Buck

The Buck Starts Here

You might be the world's greatest e-wheeler, but if you don't have a buck to work to, you're wasting your time. It is a rare panel indeed that can be formed without a buck of some type. Natural human laziness often tempts us to think that a panel is so easy to form we can just wing it and it will come out close enough. Buck, shmuck. In reality, though, that never works. Sheet metal panels must have a model to be fitted to. How much time you spend making your buck determines how accurate and detailed your panel will be. It is very common to spend more time making

The original 1956 wooden buck from Dante Giacosa' blueprints for the Fiat 500 as seen in Fiat's museum in Turin.

the buck than the actual panel. Read that sentence again and take it to heart. You simply can't join panels together in theoretical space. You've got to have a tangible contour along which they join. The good news is that new ideas, and new technology, have made creating bucks a lot easier than in the past.

TRADITIONAL SOLID BUCKS

Solid bucks made from wood, or clay have been around for several thousand years. Gold and silversmiths used them to hammer against to form objects. Fine detail could be cut into them, and then a chasing tool, today often called a "corking" tool (a mispronunciation of caulking tool, commonly used by shipwrights), is used to hammer in all the detail lines. Though less common today, they are still used, even by high-end carrozzeria like Zagato in Italy.

EGG CRATE BUCKS

Probably the first thing you think of when someone mentions a buck is the classic egg crate, an arrangement of different wooden profiles fitted at 90 degrees to each other. They've been around for one hundred years because they are simple to lay out, logical to build, easy to modify, and they work. Really useful egg-crate bucks have solid areas attached to the profile sections upon which the panel beater can…literally…beat his panel thus forming difficult shapes in exactly the right spot. Such

Notice how large sections of the buck are detachable thus simplifying design modifications.

Stashed in a backroom of the America on Wheels museum in Allentown, Pennsylvania are Hiram Hillegass' underappreciated cast iron sprint car bucks from the 1940s. Hillegass, more than anyone else, brought affordable racing to the common man. Talk about curatorial malfeasance!

Hillegass used connections at Bethlehem Steel to get bucks cast. He then quickly formed aluminum skins around them which he sold to budding racers. The dimples in the buck show where the separate panels were joined together.

Back in the 1970s it was common to replace dented or rusted sheet metal panels with non-factory fiberglass parts (yuk!) 'Glass panels can redeem themselves by becoming bucks as with this Jag E-type bonnet.

Here's the alloy bonnet made using the 'glass buck. The builder cut the wheeled panels down 3mm before welding to keep the original dimensions of the buck. It fits the car perfectly.

areas are usually found at the grille opening, wheel arches, and window frames.

Many adherents to the egg crate save money by only building a half the buck. Once they form the panels for one half of the project, they flip the profiles around mirror-image fashion and then form the other half. This "great idea" doesn't always work because it sometimes is more cost effective to build panels using the entire span of the buck. Flipping a buck the other way makes it harder to do final welding on the first side, too. I've done it, but try to avoid it in most cases.

One interesting phenomenon connected with egg crates is that people don't like to get rid of them once the project is completed. Egg crates quite often get stuffed in a loft, stored in an out building, or even become furniture, or home décor. One well-heeled gent in Colorado had a master fabricator build him a custom Bugatti body. He loved the buck as much as the finished car and wound up having the wood varnished and the piece hung from the cathedral ceiling in his living room. Now that's a car guy after my own heart!

Italian Wire Bucks

"Wire" bucks (actually steel rod and other mill shapes) were popular in Italy, but much less so elsewhere. Forming these required a good eye and a lot of oxy-acetylene gas because many of the pieces had to be hot formed, and all the pieces had to be welded together. The real benefit of a wire buck is that it allows for unsurpassed access to both sides of a panel, a feature that is very important in getting a panel to lie down accurately, and sometimes lacking in poorly designed egg crate bucks. Good wire bucks also incorporate solid areas at key points to assist panel beating.

Bondo and Cement Bucks

An easy way to make a buck is to simply pour concrete into a hollow part, or impress the part into concrete. In both cases use a parting agent, of course, like PAM, Vaseline, or a thin skim of axle grease. The best type of concrete mix to use is fiberglass reinforced. Make sure you are scrupulous about mixing it with the right amount of water. Contrary to what many think, concrete does not

Carrozzeria Bertone built this lovely egg crate of their 1954 design, the Alfa-Romeo Giullieta Sprint.

The buck and this 1958 Coventry Climax powered Tojiero were snapped at Wayne Carini's Portland, Connecticut shop. Gorgeous!

Notice the solid wood carvings around the window, tail lights, and wheel arches to assist the panel beaters.

Mark Barton cuts an access hole in this Maserati buck he's building.

The manufacturer is long gone, but this buck for John Tojiero's little race cars is too precious to dispose of. Access holes are highly important to the utility of an egg crate buck.

He uses these home-made non-marring wood clamps to hold the panels in place.

Craig Peterson of Dagger Tools teaches metal shaping classes. Here he demonstrates how to use a welding rod to fair the profiles on a buck. Note the metal bracket that forms the wheel arch opening. It makes turning the flange for the wire edge much easier.

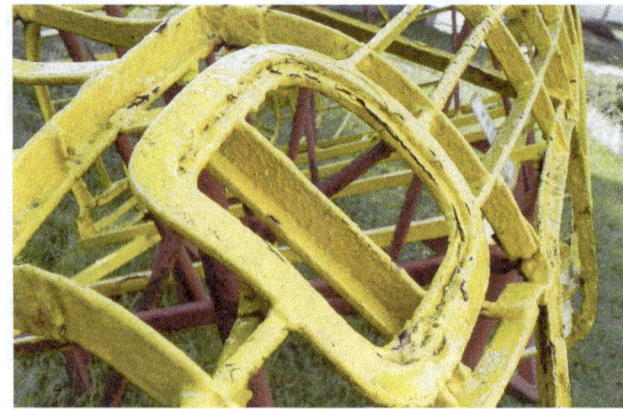

Detail of the gas filler inset shows that this wire buck had a lot of time lavished on it because it was meant to be a hammer form for actual bodies. The painted over rust is proof of yet more curatorial malfeasance.

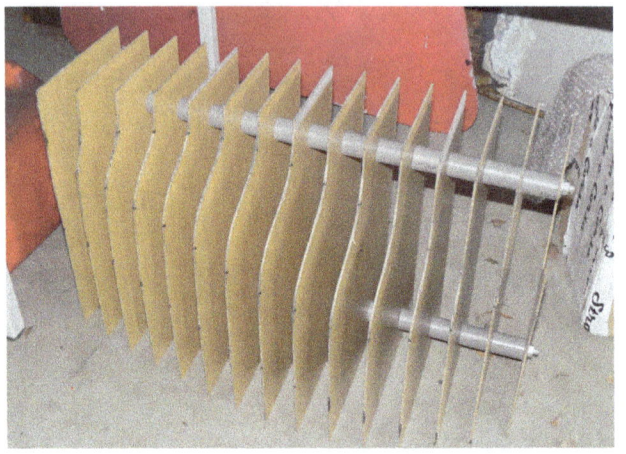

Ben van Berlo's variation on an egg crate. By joining the panels with round tubes he can quickly flip them over to build both left and right side Porsche 356 inner wheel wells.

The Stanguellini Bialbero 1100 wire buck in Modena.

Whose bright idea was it to leave this all-steel buck for a 1958 Lancia Appia GTE Zagato outside in the rain and snow at the Museo dell'Automobile near Basano del Grappa in northern Italy?

A race ravaged Bialbero awaiting a new body in Connecticut. Most likely the old body will serve as the buck for the new one.

"dry", rather it "cures". The water molecules are broken down in the chemical reaction that takes place and much of it becomes part of the concrete. The best practice is to give a concrete buck several weeks to cure before using it. It actually gets harder over time. It took the Hoover Dam 75 years to cure! Delmar Benjamin, who built the fabulous full-size flying replica of the Gee-Bee R-2 in the early 1990s used a concrete female buck into which he flow-formed aluminum sheets to create the big wheel spats he needed.

Smaller parts can be turned into a buck using nothing more than Bondo, however. Wray Schelin has made much larger parts using a combination of a steel wire inner form and Bondo. He coats the inside of the original part with paste wax. This is a fast way to capture an exact shape.

THE REMARKABLE SCHELIN FLEXIBLE SHAPE PATTERN

Hands down, the single best type of buck described in this chapter is the flexible shape pattern devised by that savant of sheet metal, Wray Schelin. I feel certain that half the people reading this book will skip this section because it's too good to be true. For the rest of you, trust me when I say that if you have a part, or an entire vehicle, that you can pull a FSP off of then you should do so because that is the best way to create a buck. FSPs are the closest thing to laser-measuring a car you can get without going through the expense and trouble of finding someone with the high tech hardware. Like many, I was skeptical of this process when I first read about it a decade ago. How could a couple of rolls of tape create a buck? Over the years, I met more and more folks who had tried it, and swore by it. Finally, a few years ago I attended a Metal Meet and saw some FSPs in the flesh and was sold. They really work!

A Schelin Flexible Shape Pattern could best be described as a female buck. The shape of the object is captured inside the body of the pattern, and it is captured so precisely that the pattern could be balled up, stuffed in an envelope, shipped across country, and the recipient would be able to make a precise part using it. It's been done.

1 The Remarkable Schelin Flexible Shape Pattern: Jordan Scowary started with cardboard as he developed the lines for his custom chopper tank.

2 He then made a metal profile frame using Wray Schelin's modified arbor press to bend the ½ inch cold rolled flat bar.

3 The 1/2 inch x 1/8 inch stock is quite rigid when welded together.

4 He then infilled the frame with Styrofoam cut on a bandsaw. Use Elmer's white glue, or spray contact cement to stick it together.

5 After the glue dried Jordan profiled the foam with blades and rasps.

6 Finally he covers the buck with J-Mac "Classic Clay" brand modeling clay. You can microwave this to soften it, but wear eye protection because it can pop suddenly when withdrawn from the oven.

TEMPLATES AND SWEEPS

Some would-be panel beaters fail to achieve the results they want because they try to make a panel using only a male buck. Templates and sweeps are required to check a panel's flow especially when it is difficult to get access behind a panel. On FSPs, templates are required because an FSP can only tell you the "area" of a panel, not its "arrangement." (More about this later.)

LASER MEASURED AND LASER CUT

Computer wizardry has brought us to the point that buck making is utterly simple if you have a big enough check book. It is now possible for a car body to be completely laser scanned and turned into a file, that file be sent anywhere in the world to a CNC cutting program, and all the pieces for a perfectly fitting egg crate buck cut out of any material the customer indicates. Plywood is still the standard although MDF board and aluminum are choices. The cost for this service is about $15,000 (11,000 Euro), but that is a bargain compared to the cost of paying a restoration shop by the hour to measure a car body, calculate the data points needed, and then rough cut and fine trim individual panels.

PANEL PLANNING

Some people are into self-flagellation. If something doesn't hurt it can't be worthwhile. That's not me. I like to get a job done as painlessly as possible and move on to the next one. Stress sucks. Learning how to divide a project into manageable panels is an important part of body fabrication. More easy-to-build panels are usually faster to form, and join, than fewer, harder-to-form parts. Ron Fournier told me that he tries to never have two highly shaped features on a single panel. If he can split a panel in two pieces at a moderately shaped area and just put one complex feature in each half, he'll save time and frustration. The more classic cars I examine, the more I find that even the great craftsmen in the days gone by adopted that practice, too. Don't try to roll a beautiful body out of one piece of metal. Use ten, or twenty, or a hundred smaller panels. Keep each panel as simple as possible, and weld them together. Let's look at some examples.

7 FSPs only work if you have an actual part as a model. Begin by laying down strips of transfer tape. This is a low-adhesion translucent tape used by vinyl lettering industry.

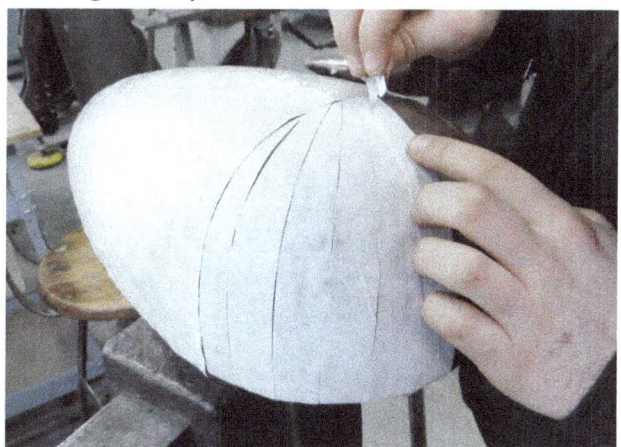

8 Do not overlap the transfer tape anywhere. Use a razor to trim as necessary.

9 Once you've covered the part with transfer tape, apply fiberglass reinforced shipping tape at 90 degrees to the transfer tape. Overlap shipping tape upon itself about an 1/8th of an inch.

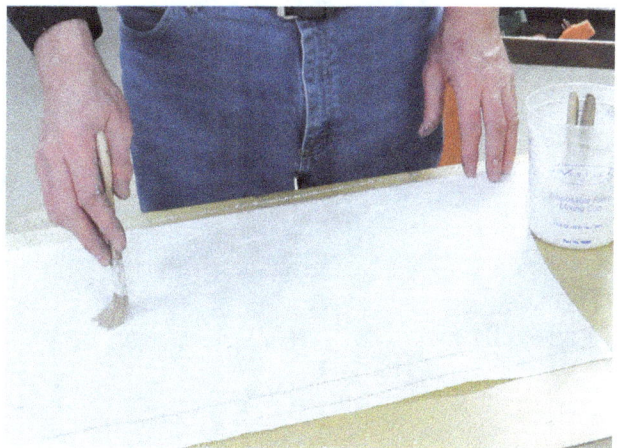

10 Carefully remove the FSP from the part and dust what little adhesive remains on the transfer tape side with plaster of paris to kill it.

11 Your FSP now needs to be "filled" by a shaped part. When the part you are shaping fills out the FSP and there are no more air gaps, your part is complete, though it may need final "arranging" (more later).

12 Schelin stores several complete car FSP collections in his "library". How much room would egg crate bucks take up?

13 Jordan marks on his FSP the lines where each template should fit.

14 Highly accurate templates can be made out of 19 gauge steel sheared ½ inch wide and then curved in a shrinker-stretcher.

15 Wray's monster arbor press can exert a lot of force, but the ½" (13mm) steel templates Jordan creates must fit the buck to within 1/32" (1mm) along their whole length. Wray has used this machine to build entire car bucks out of steel.

An FSP was all Anders Nørgaard of Denmark needed to replicate this Porsche 356 transmission inspection plate on the left.

Though Mark Stuart built a wooden buck, seen here in the early stages, he found it easier to wheel up panels for his replica Daytona Cobra using FSPs. The results were impressive.

Wanna make a fast buck? Try using aluminum mesh as seen here at a German car show. Once you remove the mesh cover it with fiberglass to retain its shape.

Jessy and Amanda Blandford, students at a Dagger Tools class, tack a paper pattern onto a buck as they begin to divide their fender into logical segments.

The Schelin's "Bondo buck" system uses metal framework embedded in body filler to capture a shape. Randy Ferguson shows one he made from a '40 Willys trunk.

Only the use of a different alloy in the filler rod gives away this perfectly finished weld line on a 1938 Alvis 12/70 cowl.

The body filler conformed perfectly to the interior of the Jag 140 fender and is kept from fracturing by the welded wire frame. Cheap, easy, and fast.

Typical division of a '30s style fender

Even very small radii need a template. Here Wray Schelin makes one from paper for a 1954 Buick.

Steve Hall creates the template from the undamaged left side of a race car so he can later rebuild the wrecked side to the same profile.

A Bugatti "Bordino" tail from the outside.

The front lower nose for a Bugatti rear fender where it meets the running board. These apparently simple pieces are made from three welded pieces.

Even the "simple" clam shell bonnet of a Type 35 Bugatti separates into 4 pieces, 2 panels per side. The orange tabs show the weld lines. This one is at the Simeone Foundation Museum in Philadelphia.

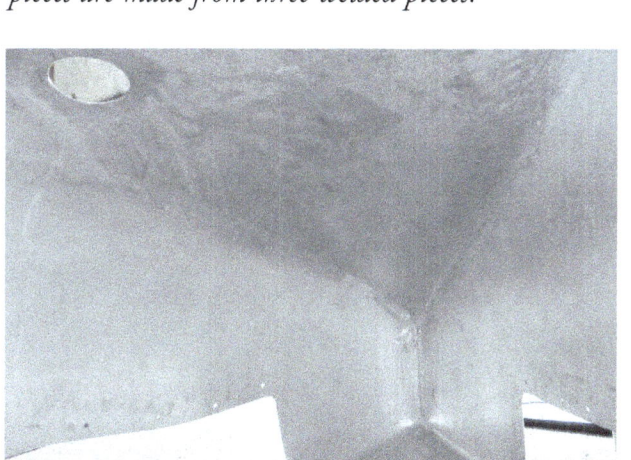

A Bugatti "Bordino" tail cone from the inside. You can see the welds where the four panels join.

Ben van Berlo indicates one of the five weld lines that join this Porsche 550 Spyder rear panel.

Some of the many panel join lines on the front of a Lancia Aurelia.

Chapter Seven

Birdcage Restoration - Part 1

Building the Bonnet

Mark Barton's build-up of the bonnet for an ultra-rare 1962 Maserati Tipo 151, a car worth over $15 million, took longer than my time with him, but you can see the most important stages of the process in this sequence. The goal was to return the car back to its original mid-1960s form which meant narrowing the wheel wells and re-profiling the engine cover. What impressed me most about Mark was not simply the level of craftsmanship, but his speed and the intensity of his focus when wheeling. Again and again I found that deadlines mean something at The Panel Shop.

This car was scheduled to go racing shortly after this photo was taken. The Panel Shop meets yet another impossible deadline!

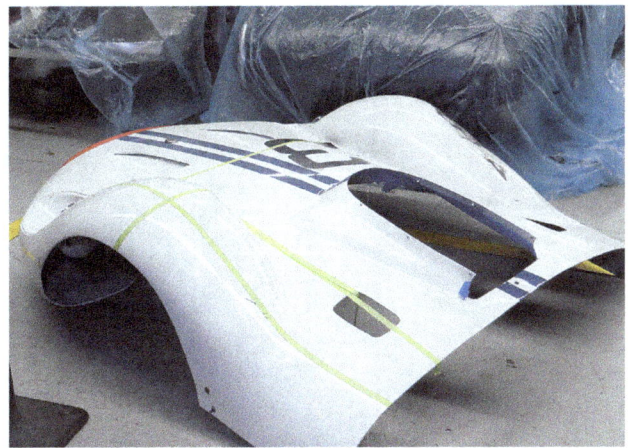

The pranged "Birdcage" bonnet whose wheel wells had been widened over the years as tires evolved.

Here Mark presses just a bit of shape into this panel prior to wheeling. The slight shape he induces by dragging it across his chest prevents the panel from flopping in the e-wheel during the early stages.

The classic egg crate buck. Note the bent ¼ inch rod over which the flanges will be bent. This buck was designed to be disassembled and reversed to shape the right side panels.

Frost e-wheel uses anvils with flats cut on them. Note screw on the right side of the axle used to adjust anvil/upper wheel alignment. RIGHT: Using the #2 anvil and close tracking Mark puts shape in the panel.

Mark lays out a panel blank to be cut on the stomp shear. The white material is a plastic protective layer on one side of the 3003-14 aluminum which must be removed prior to wheeling.

The panel is actually on the inside of the wheel well. Open structure of the egg crate allows him to see where more wheeling is needed to settle it down. Wherever the panel touches the buck it should be wheeled.

In less than an hour the panel hits all three plywood stations. Don't try to be this fast when just starting out!

Wherever your panel touches the buck, you wheel the panel more. Most of Mark's wheeling is done vertically (up and down on the panel) using a flat wheel.

This panel had several shapes: two distinct areas along the top and one on the bottom. Less proficient wheelers might choose to make this in 2 pieces. Mark's experience taught him to make fewer panels, faster.

Most panels require their edges to be shrunk, so maintain your no-blow-zone. Here Mark does more shrinking using a WWII era Erco kick shrinker.

Having your e-wheel literally one step away from the buck saves time, but also improves build quality because you will be more inclined to check your progress more often.

On this tight radius he pulls the panel down slightly with each pass.

The tracking marks tell the story: Mostly vertical wheeling.

Sean aids his father, notice how he keeps his fingers well away from the pinch point. RIGHT: Forming the wheel bump from 4 separate panels is easier, faster, and more accurate than trying to wheel, or block hammer, a high crown into one panel.

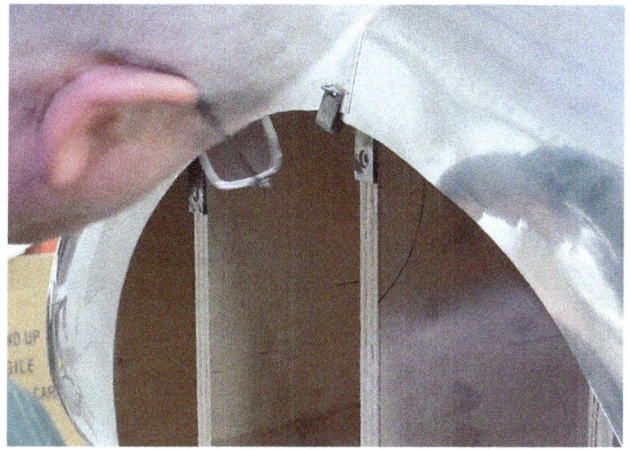

Though Mark forms panels rapidly, he never uses an overly radiused anvil to speed the process. Only an appropriately flat anvil and close tracking can give him this level of mirror finish.

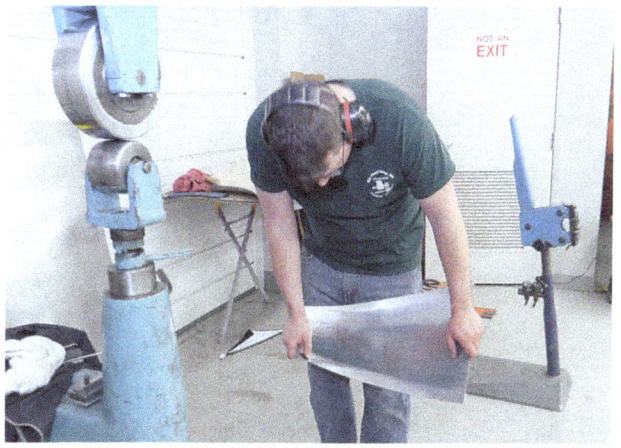

Wheeling often induces twists into panels. Once you find you have a twist (by checking it against the buck) use hand manipulation to untwist it before you continue to wheel.

One lone American Cleco lies next to a gaggle of German Würth welding clips, and their installation tool. The clips are hard to get in the U.S. but well worth the search.

Once again, most edges need to be shrunk so Sean puts the Eckold into action.

Mark cuts out a cardboard pattern that follows the contours of the bulges of the engine cam covers. Forming those bulges will take mallet, slapper, and wheel work.

His left hand pushes gently, but firmly down on the panel while his right hand drives the panel through the pinch point along the flange line.

Note the folded up shim of paper under the right side axel, used to create a pinch point between the wheels so a flange can be turned. Without the pinch you get unwanted stretch in the adjoining panel.

The accuracy is excellent. Notice that the upper wheel has been chrome plated for additional hardness.

Mark uses a Sharpie to layout his flange line.

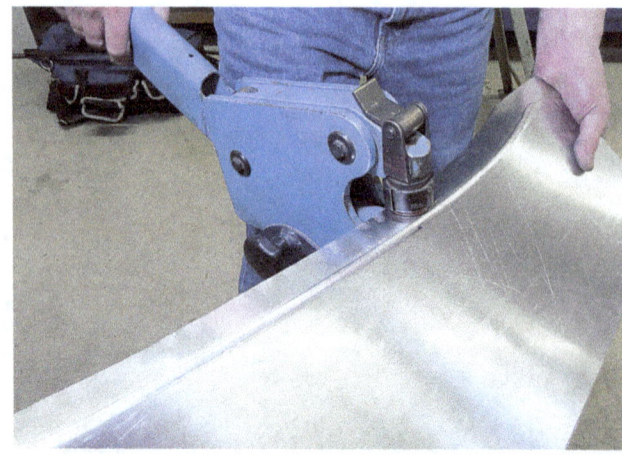

Once the flange is turned he uses the Eckold to tighten the radius further.

A little wheeling on the panel longitudinally gives it a slight curve. Notice the scallop marks on the next panel in front of the cam covers. It was crucial to create a wooden engine buck to get the panel fit exact.

First, Mark throws in a few sideway passes to curve the panel laterally.

28 To further refine the flange, Mark makes a few passes through the e-wheel. By stretching the flange a little more in this manner, it will lie over more and meet the next panel above it.

He next puts in many more longitudinal passes to bend the panel down towards the grille opening.

Preparing the front engine cover panel for flanging.

He supports the panel from the sides so it doesn't flop during wheeling. Gravity can effect the shape of a panel, as can the strength of an operator's arms - if one arm is stronger than the other, you'll twist the panel.

Once the flange is turned the panel will be very stiff laterally, so it is important to get the overall shape right before flanging.

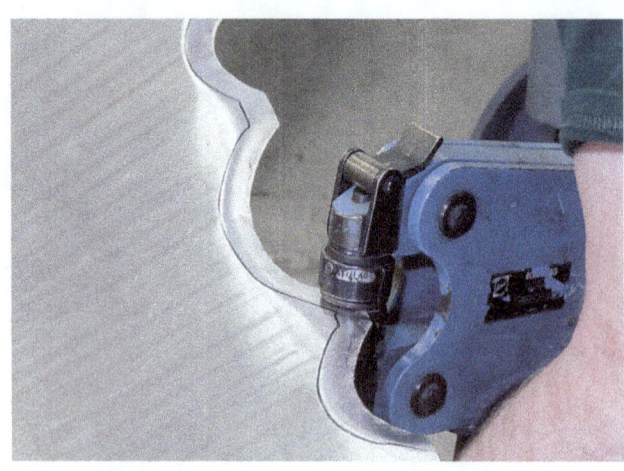

The Eckold is your friend when it comes to tight corners.

Again, he uses a paper shim to create a pinch point before turning the flange.

Here Mark doesn't have much more than 30 minutes in the panel. Eventually, the flanges from these 3 panels will joint precisely for welding. The box above the engine mockup represents the velocity stacks and cover.

Look how tight and exact this scalloped flange is. It was done without a rotary machine at all!

The next panel begins with a simple fold biased to its left side. Note the impact damage on the bonnet. Though it could have been saved, the bonnet had already been modified from its original shape.

After forming the fold Marks uses a rounder anvil to begin to roll the fold. RIGHT: A test fit shows that the panel needs more sweep fore and aft, and where the front cam bulge needs to be formed.

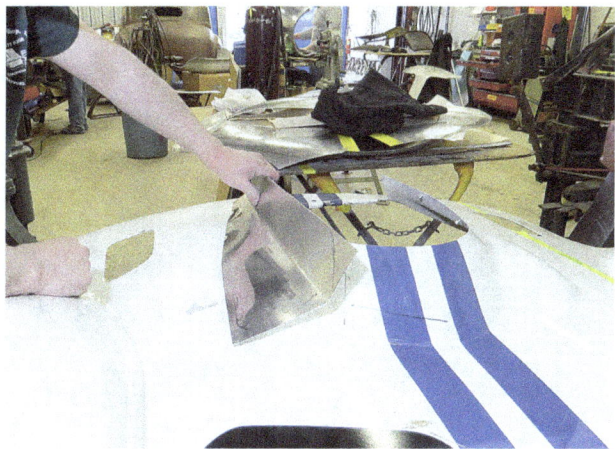

Through hand manipulation, the panel begins to conform to the raised bonnet cowl.

To create more sweep, Mark continues to stretch the fold in the wheel.

Using a No. 2 wheel, Mark smooths out the panel and blends the angular twists.

Notice the waviness of the horizontal fold. This means he has enough extra material now to force the panel down fore and aft to conform to the bonnet curve.

After initial trimming, the panel fits into position well. The next step is to form the cam bulge.

To shape the metal closely around the cam bulge, Mark anneals the end of the panel with a carburizing flame - one that leaves black soot on the surface, the soot will be burned off next with a neutral flame.

More smoothing of the bulge. With low pressure, slow passes and a close track.

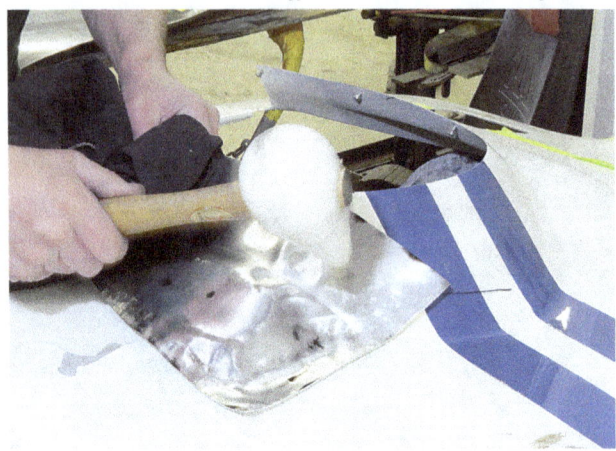

The annealed aluminum responds very well to the pointed end of a plastic teardrop mallet.

The Eckold is used here to stretch the upturned flange. Changing jaws in this Swiss-built machine is a snap.

The walnuts formed by the teardrop mallet are planished in the wheel.

There is a lot of shape in this small panel already, yet the process is simple, logical, and requires only a few tools.

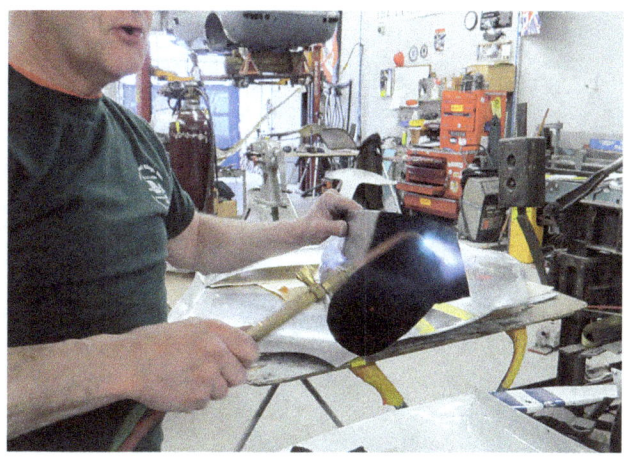

Refining the shape requires more annealing. Mark uses a neutral flame to burn off the soot from the carburizing flame. Carbon burns off at 750 degrees, the temperature needed to anneal (soften) aluminum.

A test fit on the buck. Final flange work will be done when all the panels are finished.

Mark uses a slapper ('flipper' according to him) to coax the aluminum into conformity.

Measuring for the next panel.

More slapper/flipper work, this time on a cast forming head. The goal now is to planish the part to give it a smooth finish.

This panel, which sits just above the grille, has a flange already turned on the left side and thus is very stiff. Mark still gives it ample support as he moves it through the wheel.

The panel fits well along the length of the bonnet, the curve of the wheel well, and along the grille opening. Empty space to the right of the grille was left as the final brake duct positions have not been determined.

The classic profile of a "Birdcage" bonnet emerges. "Birdcage" actually referred to the car's proliferation of chassis tubes. The panel to the right has a serious reverse wheeled into the front of it.

The grille oval was formed from very soft aluminum air conditioning tubing. The panels will be annealed before being hammered over it.

Yes, tacking and welding on a wooden buck is permissible. Some panel beaters cover their wood with tin foil, though this doesn't do a whole lot of good in preventing burns.

Mark is working right to left on the buck because once the right side is done, he'll flip some of those wood stations to create the left side buck, thus saving his client money.

Mark throws in tacks every inch using a Purox W-200 torch, the one I prefer. My next book will cover this process is much greater detail.

If your panels begin to separate at all during tacking, stop and get them back into perfect alignment before moving on.

Many panel beaters like Mark use cobalt blue glasses to weld aluminum. Ron Fournier sells these. Can you spot the three types of clamping devices used? Each has a purpose.

If you can't control them on the buck, work the seams on your bench with hammer and dolly. RIGHT: You've spent hundreds of hours wheeling a car body, so don't spoil it by rushing through the welding.

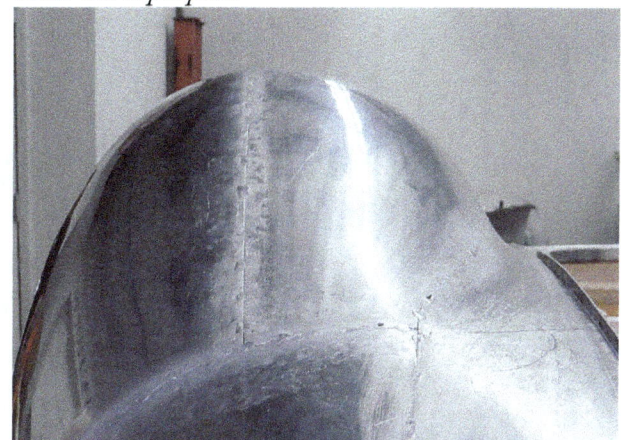

Enjoy the beauty of this shot.

Final trimming is done during the weld process, not before it. Alignment issues always occur, so wait before cutting.

Clean the flux out of a weld joint before wheeling it. Flux grit can ruin polished wheels instantly.

Mark checks the symmetry of the front bonnet panels. Little wonder that he and Steve continue to attract high end clients.

Engine cover panels are formed and ready for final welding. The yellow tape hides a transparent acrylic molding that, when heated, was draped over the aluminum male mold seen in an earlier photo.

The front half of the bonnet has been slid forward a few inches in this shot. You can see the grille surround is partially hammered.

A few weeks after the last shot, the bonnet has been completed and painted, though the engine cover is still being worked.

Mark Barton & Steve Hall

The two poles of ultra-high end classic car restoration in the U.S. are southern California and western Connecticut, just outside New York City. The amount of money available for restoration work near the Big Apple is seemingly limitless, but you'd better believe the customer expectations are as high and rigid as the Empire State Building. In the land of Wall Street billions the competition for commissions is utterly fierce. As the song goes, "If you can make it there you can make it anywhere."

Mark Barton and Steve Hall are two British blokes who have "made it there." The co-owners of the Stratford, Connecticut firm The Panel Shop came to America in the mid-1970s as refugees from England's imploding luxury, low-production cottage car industry. At 16 Mark had started an apprenticeship program with Aston-Martin while at the same age Steve entered Rolls-Royce's program. Within a few years overly-zealous unions and high taxes had nearly put paid to those companies so Mark and Steve began to look for other opportunities.

Such an opportunity came from Don Hart who, in the late 1970s, brought them to Connecticut to wheel panels for his better off-forgotten kitsch retro-speedster project called the Doval. When that pipe dream fizzled, none other than the legendary Robert Cumberford, car designer extraordinaire, who was trying to launch his better-conceived retro project called the Martinique, hired them. Cumberford had learned it was a lot cheaper to pay two lads from England to fab custom panels than create expensive stamping dies at union shops in New England.

Their efforts were not enough to save Cumberford however, whose Martinque dream ended when investors fled in the wake of the DeLorean scandal. Mark and Steve decided to try their luck in the U.S. rather than return to the fiasco of England's economy in the late 70s. They picked up piece work for a few years, but by 1984 had earned the respect of enough of the New York metropolitan's most noted car restoration shops for their high quality wheeled panels that they launched The Panel Shop. It grew from strength to strength due to a reputation for premium work, reasonable pricing, and a consistent no-drama ability to meet deadlines. Their customer base now ranges from pay-as-you-go individual car restorers of modest means, to the rich, the famous, and even the über-rich and famous.

77

Chapter Seven

Birdcage Restoration - Part 2

A Re-Roof Job

The first order of business was to build a buck to repair the partially smashed nose. Using what was left of the actual car, as well as historic photographs, a wooden buck was built upon which Mark was tasked to wheel up new panels. While he did that, Steve tackled the side of the car that had contacted the wall. He began by forming a new roof panel to replace the very corroded, modified, and thinned-out original. Before he began, he asked me how I thought the panel should be formed. Looking at it,

Here the roof panel is test fitted to the original straightened steel frame. New coating technology will ensure there is no longer any electrolytic interaction between the dissimilar metals.

I suggested from one piece cut into an "H" section and then wheel some crown into each leg of the "H". He shook my ideas off and showed me the panel had originally been formed from four individual pieces, and that if it were formed any other way it would be much more difficult to shape correctly, and nearly impossible to fit to the frame. Because each piece had multiple flanges, and the over-all shape was so spindly, it was necessary to form four pieces and weld them together.

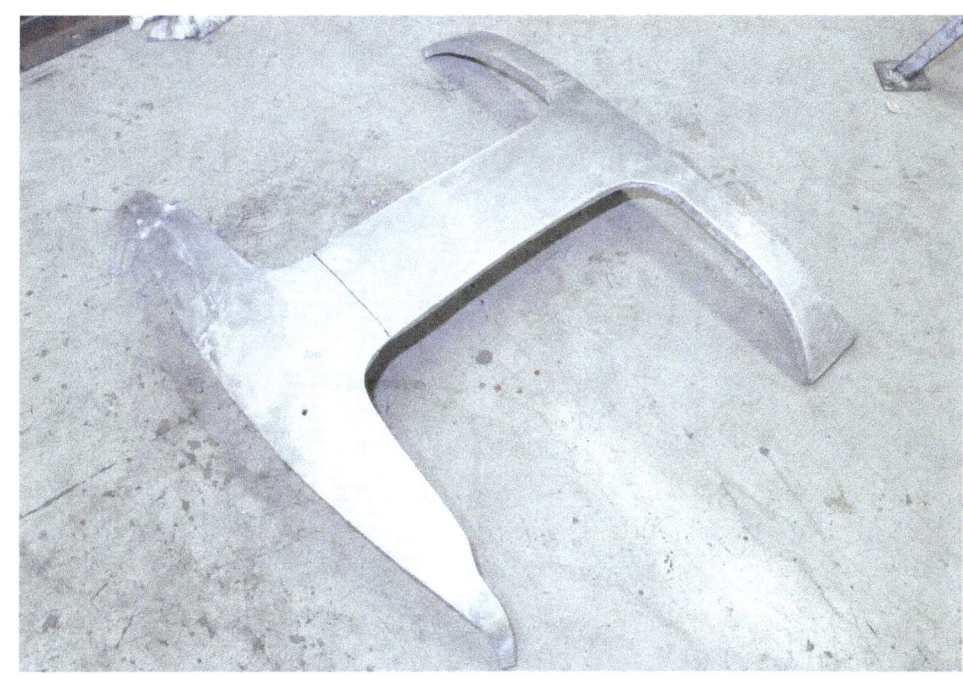

The original panel had been modified over the years, was warped due to race damage, and thin in areas where previous carozzeria had over-sanded it during repairs. Electrolytic damage due to the interaction of the aluminum panel on the steel frame of the chassis was apparent under the panel.

Steve holds an original door up to the roof panel.

You can see how 50 years of race damage and inadequate repairs have caused a gap to develop at the corner.

79

The correct repair is to form a new panel out of four pieces as per the original. Like most British panel beaters, Steve uses Gilbows to slice the "ally."

It would be impossible to wheel shape into the roof panel if it were one piece. By making it from four small pieces, as per original, each panel is easy to shape.

The new piece is clamped under the original part so that dimensions can be transferred directly.

It only took a little wheeling to put a gentle crown in this leg of the "H."

Dividers scribe a flange line all around the transferred shape. Steve's ever-present blue wrist band announces to the world he is a Chelsea FC supporter.

Notice the flat anvil in use in The Panel Shop's Frost machine. The flat, and No. 2 anvil, are used far more than any other.

Your fearless author cranks what Steve calls a "jenny" (rotary tool) so as to begin tipping a flange on the panel. Make at least three passes to develop 90 degrees. Don't do it all in one go.

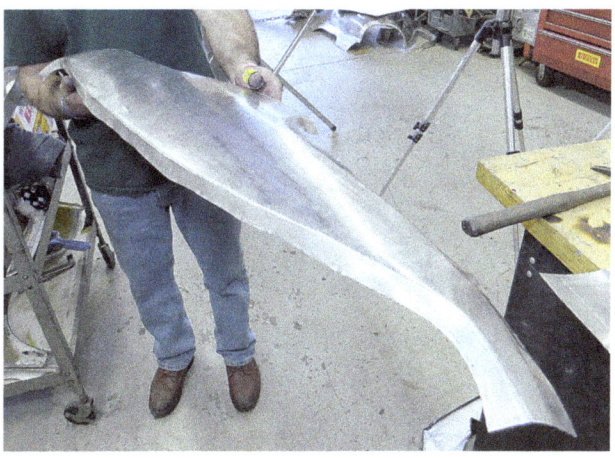

The front piece of the panel is formed exactly as the other three: wheeling, flanging, and then some judicious shrinking along the flanges with the Eckold.

The end flange is turned easily in a brake.

Sean Barton cranks the jenny as Steve gently lifts up and bends in the panel to bring the flange around.

A test fit on the original panel shows perfect alignment. Note how the door flange has not yet been turned completely. That will happen after a test fit over the car's original steel frame.

This inside radius is so tight neither the Eckold nor the Erco machines could fit in it. Therefore, Steve resorts to a maple teardrop mallet to stretch the flange.

Once the radius eases, the Eckold is used because it is less likely to tear the metal than a misplaced hammer blow.

Using an 18 inch contour gauge to compare the new panel with the original.

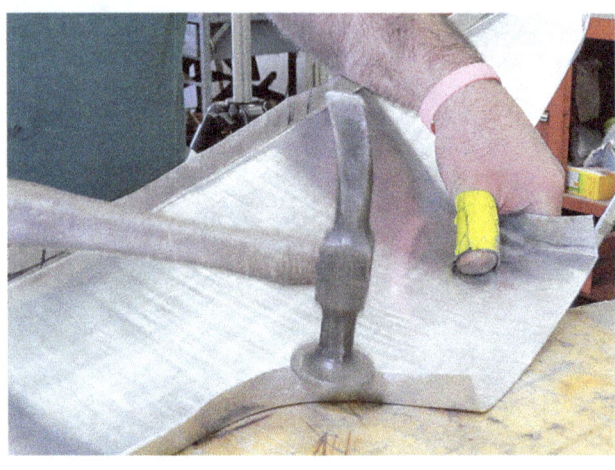

It is common for flange lines to get wavy. They're easily brought under control by hammering the flange end down onto a flat surface.

Steve finds a slight dip in the line where one leg of the "H" connects to the center piece.

Hammer and dolly work takes the waves out of the flange itself.

He corrects this quickly with some judicious mechanical stretching.

Other problem areas on the individual pieces are dealt with by light wheeling.

Not only can you see the weld lines that join the original pieces together, you can see fifty years of bad repairs and corrosion.

Another test fit shows the new panels lining up well. Steve helped build the Rolls Phantom V in the background during his apprentice days.

Gas welding is preferred because TIG welds tend to crack when crushed in an e-wheel. The orange glare is from the aluminum flux, always use special lens when welding aluminum.

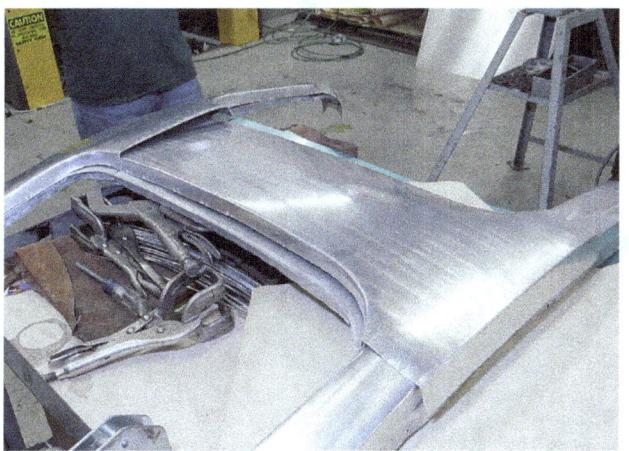

The panels are ready to be welded. Notice the tracking pattern left by the e-wheel.

The new roof panel is an exact match for the original, but isn't porous from grinding, warped from race damage, or pock-marked with corrosion. The door fit is now exact.

Chapter Eight

Jamie Downie: Back to the Future

Classic Shapes from a Young Builder

Melbourne, Australia's Jamie Downie seems to have stepped out of some kind of time warp. How else can you explain how a young Gen Xer is not only so very good working with his hands but also has the work ethic of an Outback prospector hungry to strike the mother lode? Jamie is a throwback to a hardier generation of self-supporting guildsmen and living proof that guys like this still do exist.

Panel beating is a family affair for the Downies. Jamie and Stella rebodied this 1933 Delage D8 while their son Colton watched and learned.

While a teenager, Jamie began working in a couple of traditional panel beating shops in his town. Though called an "apprentice," he wasn't enrolled in the type of formal program that had been common in Britain and Commonwealth countries. That system had mostly died out a generation earlier. Jamie started off doing menial jobs around the shops, but through observation, some instruction, and a whole lot of personal motivation he acquired metalshaping skills the way some teens of his day acquired Nintendo cartridges. At the Melbourne firms of Black Art Fabrications and The Coachbuilders he learned the extremely high standards of bespoke body building that discerning clientele expected. By the time he was 25 years of age he packed up his skills and struck out on his own starting Kustom Garage.

Kustom Garage was the classic "mom and pop" operation with Jamie's wife Stella pitching in to wheel panels, assist with the welding, and doing her bit to make the business succeed. The quality of the Downie work, and the lightning speed at which they achieved it, soon attracted a healthy customer base, and even the attention of the International Specialized Skills Institute, an Australian group dedicated to promoting the acquisition of both traditional and high tech skills by Aussies. Jamie was awarded a fellowship to travel to the U.S. to study the American style of panel fabrication, power hammers and pneumatic tools. He returned home determined to incorporate this faster style of fabrication into the traditional methods he had been practicing for some time.

Bespoke Bike Body

Jamie first came to my attention when I saw pictures of a 1930s Delage racer he had rebodied. The Arlen Ness style motorcycle body he fabbed in the following sequence only hints at the quality, variety, and artistry that is typical of Kustom Garage. You'll see a lot more of Jamie's work in my next book.

"Look on my works ye mighty and despair!" In Jamie's hands, what started as a rough sketch on a piece of paper turned into this sensuous object of metal art.

You've got to build to a buck. Notice the numerous access holes for clamps. Fair all the stations with a flexible batten to make sure they flow.

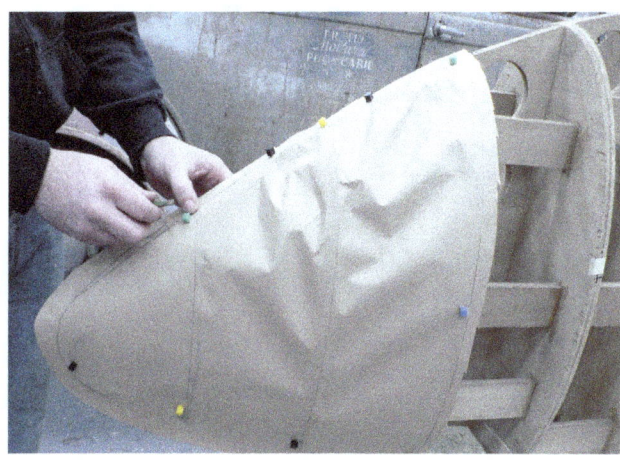

A paper pattern is used to layout the first sheet metal blank. Notice the beveling on the right most station done during the fairing process.

The earlier you define key features, the likelier you are to wheel a precision panel. Here Jamie notes the location of the weld joint for the upper and lower sections.

Lots of info in the picture: Notice the "no-blow-zone" perimeter on the panel; the use of a heavy steel blocking hammer; the sand bag on a tree stump; and, prudently, the ear muffs.

By developing all his paper patterns at the outset, Jamie is able to avoid waste on his aluminum sheet.

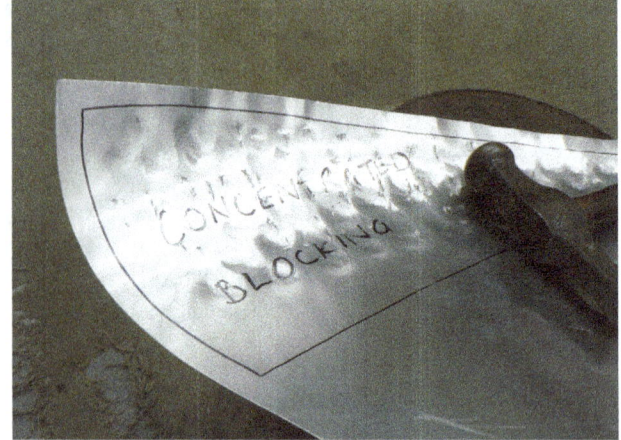

Yes, you can use a steel blocking hammer on aluminum. Ball-peen hammers are not recommended because one end is too flat, and the other too round.

Fine lines are hard to see while cutting blanks, but taped lines give the excess width and are easy to see, even for eyes older than Jamie's.

Once some shape is blocked into the panel Jamie planishes it in the e-wheel. Once smooth, he cranks in a little more pressure and continues to shape. Note the tracking.

A continual balance between blocking and wheeling gets this panel into shape quickly. Note the crossed tracking marks and their absence in the "no-blow-zone".

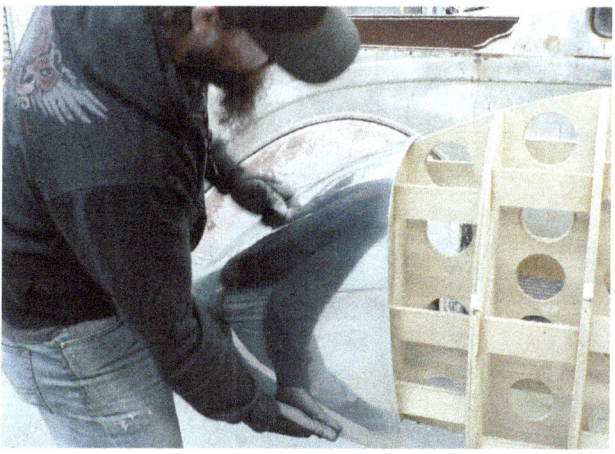

The "wasp tail" shape of this panel would cause many to make it out of two pieces split in the middle horizontally. No worries for a panel beater with Jamie's skills.

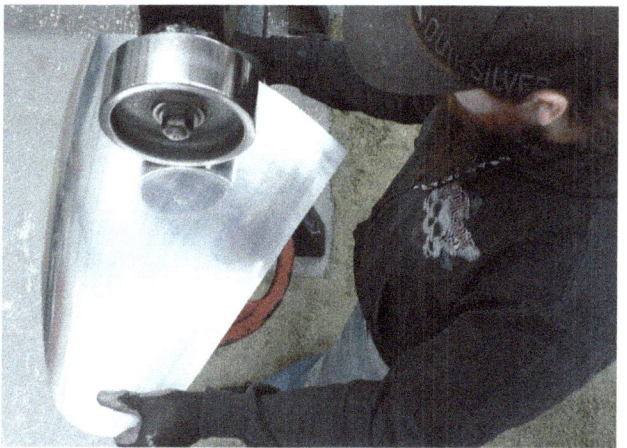

Very close tracking, light pressure, and patience gives you this shine. More and more wheelers are using thin gloves to grip the panel. Wray Schelin sells the best ones I've used.

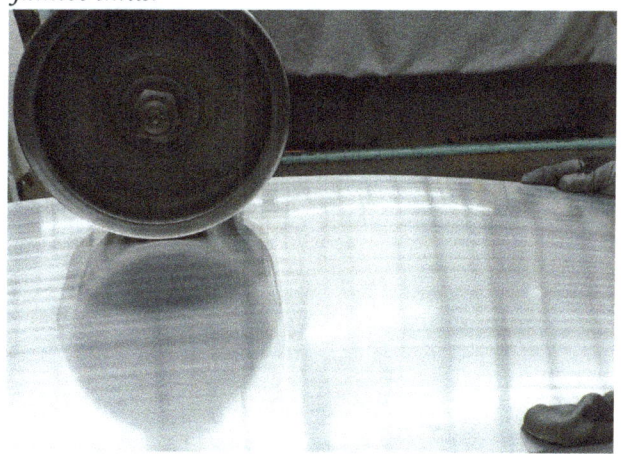

The very tight tracking evidenced in this photo sets the standard for your work. More tracks, rather than a more radiused anvil, gives you a much better panel.

The Mk I Human Knee makes an excellent forming tool when changing the arrangement of a panel. Note the green stand of the ubiquitous Eckold Handformer in the back.

Though the majority of shape was put into the panel with length-wise wheeling, by cross wheeling Jamie tightens the lateral curvature.

Three hands are better than one! Notice the blocking hammer blows are not haphazard, but laid down in well-placed rows. An e-wheel is nothing but a rotary hammer.

Planishing out the remaining walnuts requires minimal pressure. As you wheel, you'll hear the walnuts clattering as they go down. When the clatter dies off, crank in a little more pressure and keep wheeling.

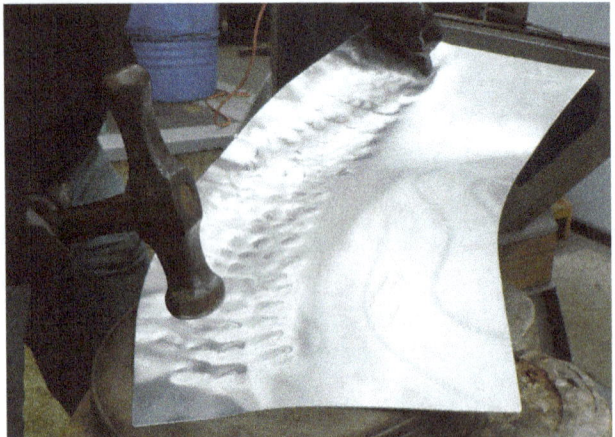

Curved rows of hammer blows kick up a curved row of "ruffles" on the left edge - this is what he wanted. Now he can use the blocking hammer lightly along the edge to close the ruffles shrinking the edge. Shrinking this way is preferable to mechanical shrinking.

The ghosted rings of the blocking hammer are still visible, but cannot be felt. Shrink needs to happen along the top and the right edges.

Most of the ruffles are gone and the edge shrink has really pulled the panel into shape.

More methodical blocking hammer blows. These big blocking hammers are hard to find. In the U.S., old Pexto No. 4s are commonly available but really too small for this kind of work.

The key to this blocking hammer is its heft and the mushroom shape of its ends.

In the previous picture the panel hit the buck in the center, so Jamie blocks more shape there. Where a panel touches a buck, it should be blocked or wheeled until the entire panel touches all stations of the buck.

When planishing in an e-wheel, be sure to continue wheeling into the adjoining areas of the panel or you'll develop a hump in the planished area.

The wheeling follows the curvature of the panel, or flat spots will develop.

The right edge of this panel must not be stretched, although about 1 inch to the right you see how the next panel will have to be stretched into a reverse to join the first panel.

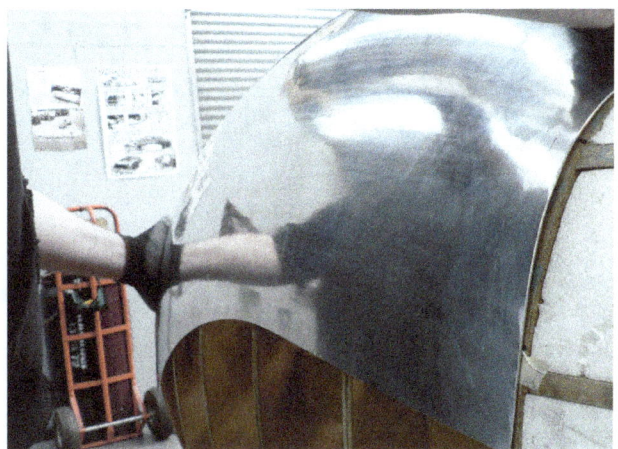

There is an organic beauty to panel work that fits this precisely.

89

Though Jamie uses a classic cast iron e-wheel I've seen work of this quality created on welded steel machines. Gorgeous!

Jamie back-shields all his TIG welds and claims they are as durable as a traditional gas weld. It's hard to argue with his success and experience.

Jamie preps the back of the panel for welding by cleaning the oxides off the surface with a Scotchbrite, acetone and then a stainless steel brush. Don't use any other kind.

The front sides of the panels are also scuffed with a stainless brush before being TIG tacked.

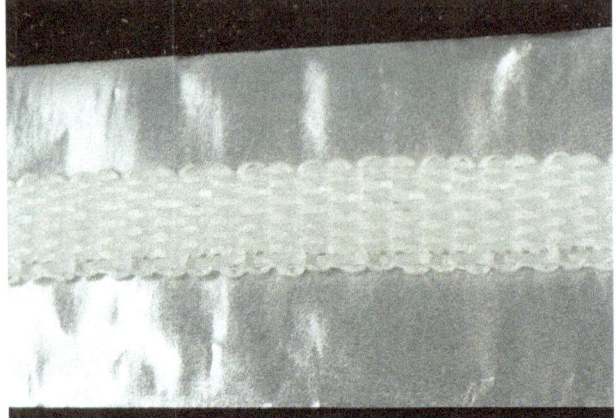

A strip of fiberglass tape, or header wrap, inside a length of aluminized tape provides an excellent backing for a TIG welded joint. The fiberglass acts to trap the argon gas, thus back-shielding the weld.

Pre-heating with a torch is a must to get the moisture out of the panel; water boils at 212 degrees Fahrenheit. Note the moisture waving out as the torch blows past.

Not quite "out of position" welding, but close. Jamie keeps his right elbow on the sandbag used to stabilize the panel. Note the big gas lens on the torch. Gas lens are well worth their small cost.

After welding, some tweaking in the e-wheel is necessary.

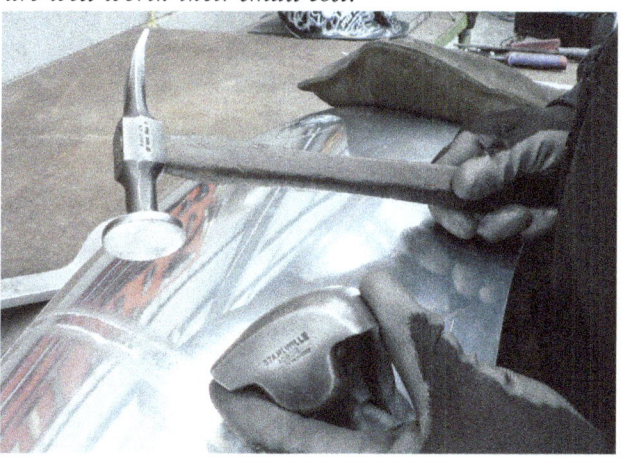

A German Stahlwille dolly and an American Snap-On body hammer unite to planish the weld line. Jamie's got such a level weld line that not much more than a light planish is needed to metal finish it.

The trickiest panel on the bike is this one under the saddle. It features a crown, a reverse, and a radiused flange that must fit exactly. The cuts on the top indicate the stretch needed for the reverse.

Using a slapper/flipper across the weld assures a smoother transition between the panels. This tool is home-made.

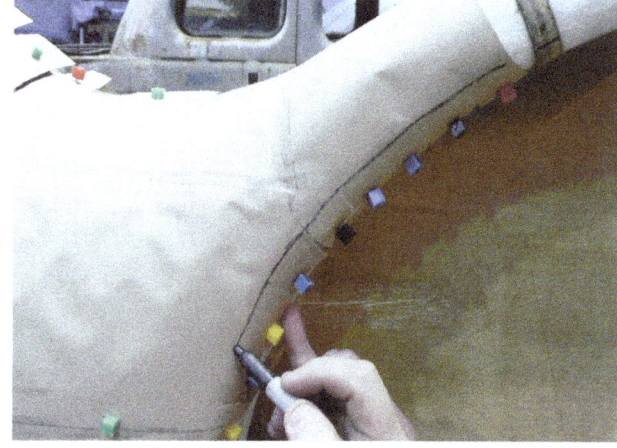

The cuts behind the row of tacks indicate the stretch needed once the flange goes over. The pen line indicates the apex of the crown.

No guts, no glory. Jamie is going to make this complex panel in one piece!

The "reverse" is begun by blocking. It is critical to stretch the edge more and fade the stretch going inboard. You can see from the hammer blows he's doing that.

He begins by wheeling vertically in the lower-rear section of the panel. Notice how he avoids the upper area (by right hand thumb) because this area will have a reverse set into it.

A long way to go, but the lower left of the panel has a crown developing while the "reverse" (sometimes called 'saddle') is well established up top.

Note the little hash marks which help him stay in bounds while raising the lower quadrant. By cross wheeling carefully here he raises the section evenly.

Typically reverses are wheeled inverted in a panel, but Jamie is using very light pressure here just to do some slight planishing.

A lot more stretching of the reverse is required, so the panel is softened through annealing. 1.) Lay on a thin soot coat with a carburizing flame 2.) Burn off the soot with a neutral flame.

He planishes the walnuts and stretches slightly along the edge. Other wheelers would "Chinese wheel" the panel inverted.

Stretching the buttery soft annealed area is now very easy.

Blending the reverse into the adjoining panel.

Being able to see under a buck is so important. The panel fits very well except along the edge which needs more stretching.

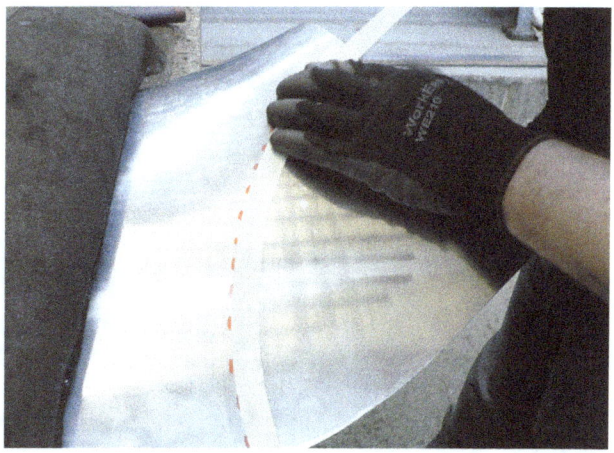

The ¾ inch (20mm) masking tape establishes the flange.

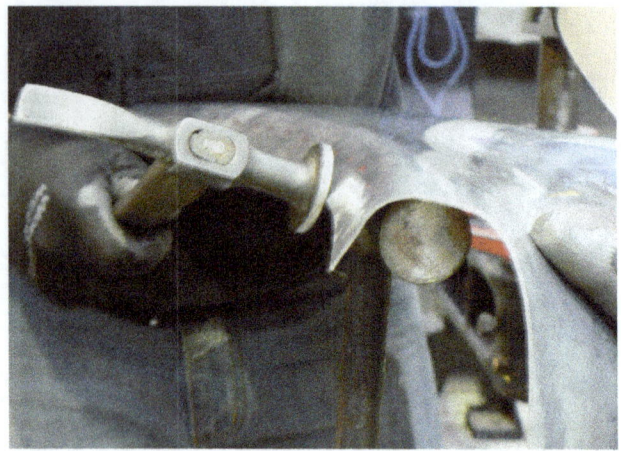

Knocking the flange over a T-dolly. He makes several passes gradually moving the metal.

At this stage, the panel work is complete. The owner could decide to polish the aluminum, or paint it.

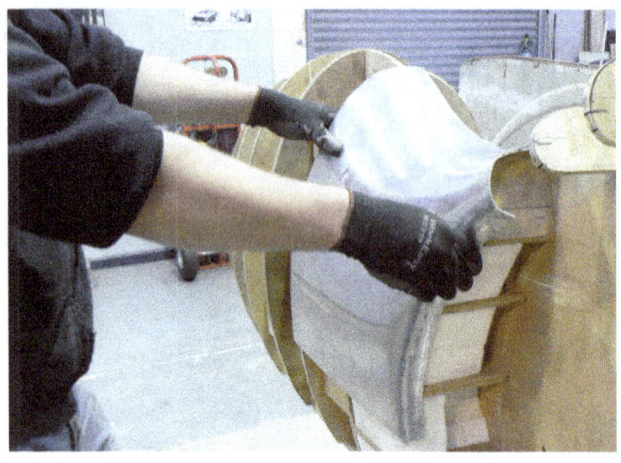

The anticlastic "saddle" shape was formed by stretching. It is one of the hardest challenges of all panel beating.

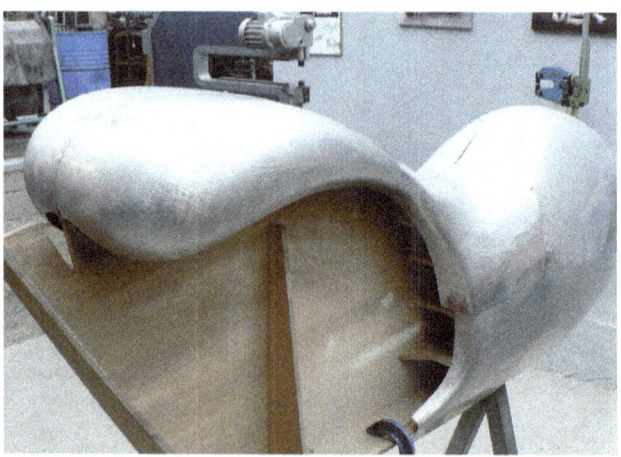

This beautiful body would have been impossible to achieve without the buck, however, bucks usually get no credit.

The transition from the wheel cover to the seat panel flows effortlessly, yet we know how much work it actually took.

After shooting some machinist's blue on the body Jamie scribes a cut and flange line with the aid of a huge milling table.

Jamie Downie Statement

As a pre apprentice learning to make my start in the trade I had no idea that I would have ended up where I am today.

When people ask why I love what I do, it's an easy question to answer but there are so many aspects to it and not just one defines why. While metal shaping is a craft that can be learnt, it takes passion, patience and the ability to problem solve along the way to take it to the advanced level. These reasons along with knowing that you are restoring history and leaving a small mark with every job done, knowing that those jobs will be around long after I'm gone for others to appreciate, and also to take a step back and look at what you have made with your hands leaves you with a total sense of achievement.

When I look back at my journey I am humbled to have met and spent the time working with all the various talented trades people, young and old who have helped shape my path to where I am today.

The art of metal shaping has become quite popular in recent years and while the younger metal shaping community is growing, the older skilled generations are dying away. I have been extremely lucky in my time to have worked with, and been mentored by, many notable artisans in both traditional British and American methods of metal shaping. From this experience I gained much knowledge and expanded my skill sets in the use of traditional hand tools and the wheeling machine to powered equipment like the pullmax and power hammers.

Starting Kustom Garage was not only a way to pursue my passion, but to make sure a dying art is kept alive and to be able to encourage and inspire others to create and restore using from simple hand tools to advanced shaping equipment. Using Kustom Garage to contribute to projects, publications and seminar's I can help make sure that knowledge is being passed on, and the trade is kept going. With this in mind we hope you enjoy and learn from our contribution to the book.

Too beautiful for words…and too beautiful for the street?

Chapter Nine

Make a Skylark Whole Again

Using a Flexible Shape Pattern

As I've pointed out before, you have to have a buck to build a panel. But what happens when a client brings you a car missing an entire quarter panel? Thanks to the flexible shape pattern, you can create a FSP buck from the remaining side of the car, turn the pattern inside out, and wheel up a panel for the missing side. If you think that's impossible just watch Wray do it in this sequence.

The '54 Skylark was one of the great understated beauties of Detroit in an era noted for excess. Photo courtesy of Martin Godbey of Vintage Motors, Sarasota, Florida.

No quarter panel, no buck, no problem. Make a flexible shape pattern from the passenger side and flip it inside out.

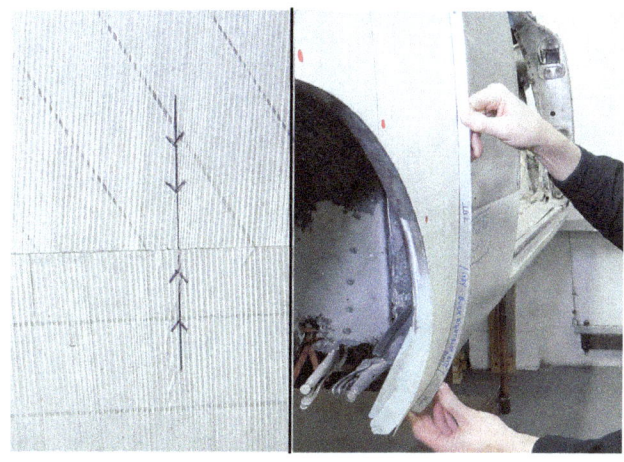

These hash marks will help Wray realign the panels prior to welding. RIGHT: ½ inch 19 gauge steel bent inch by inch in a shrinker/stretcher forms a template. Each template is numbered for location/vehicle.

Whether you're making a motorcycle tank, a car quarter panel, or rebodying a one of a kind exotic, a flexible shape pattern should be where you start.

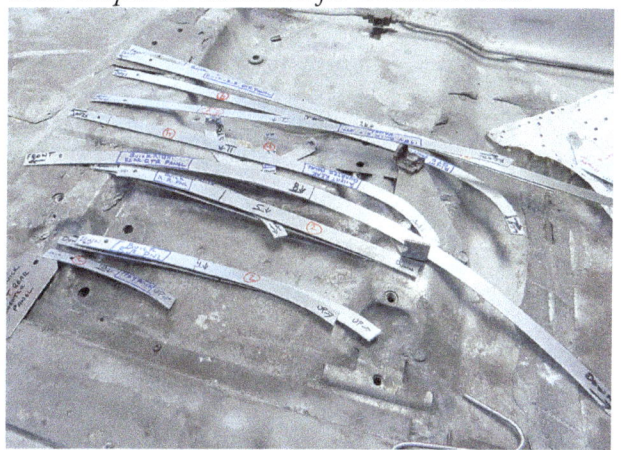

A partial collection of the templates for the Skylark. Wray sometimes uses clear tape to preserve the notations made with a Sharpie pen.

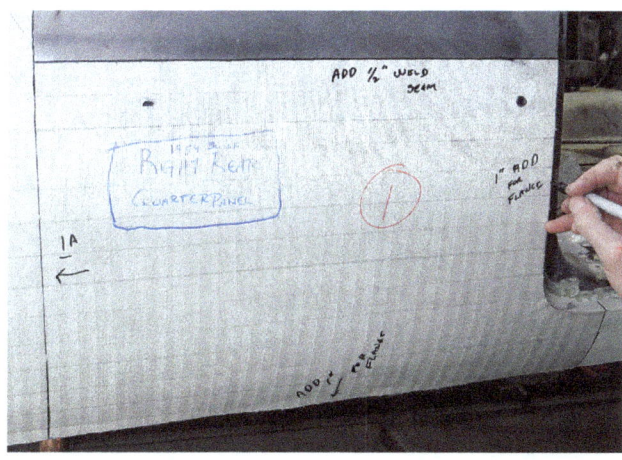

Don't remove the pattern until you've marked notations on it and made steel templates for it. The black lines show where templates line up. The black dots indicate molding holes.

The upper crescent shape along the rear of the panel demands its own panel. There is too much subtle shape here to incorporate it into the major panel to its right.

97

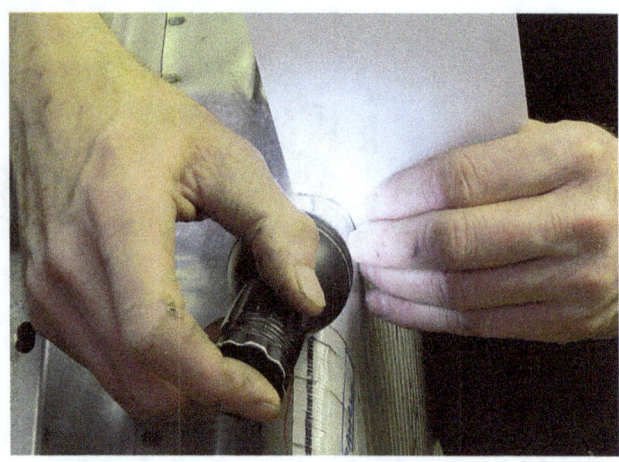

The radius is too tight for a metal template so he makes one from card stock. The light is used to check the conformity of the card to the panel. It has to be 100% accurate.

Punching holes along the template line is a must.

Wray peels off the FSP. It will retain its shape perfectly even though it is so narrow and highly curved.

Trimming the blank with B-2 Beverly Shears.

Dusting plaster of paris on the back kills residual stickiness in the transfer tape.

Marking the blank through the punched template holes. The only way to make an accurate panel is to make each sub-panel as accurately as possible.

Wray built this kick shrinker which he now uses to slightly curve the blank.

Wray used bolts with rare earth magnets epoxied to them to hold the FSP to the blank. By feeling for looseness in the FSP he knows where to wheel in more shape.

You won't see high crowned wheels used often in this book, but for this panel it is necessary.

The blank is so small it is better to pull it through the machine than risk fingers trying to push it through.

The 6 inch upper wheel on Wray's self-designed machine started life as a ball bearing used in a machine to make sandpaper.

The e-wheel has done as much as it can do so…

...Wray uses a maple slapper over a leather covered dolly to bring the edges over further.

A check against the template shows the piece is not sufficiently curved.

The radius gauge template shows that the edge needs more shrinking.

Holding the template against the blank as he shrinks the inner edge solves the problem.

Knocking the edge over further creates ruffles which are shrunk against the dolly.

Not satisfied with the slight undulations caused by the shrinker, Wray planishes them out with a machine he designed and built.

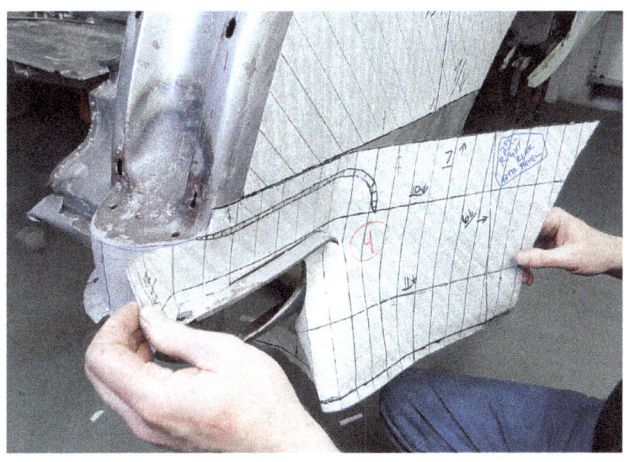

He moves on to the highly detailed lower rear panel which will be formed inverted. This is only possible because of the FSP.

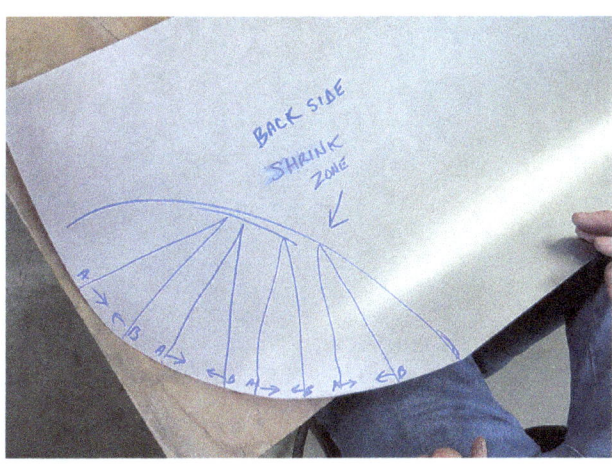

Shaping a panel inverted calls for careful planning.

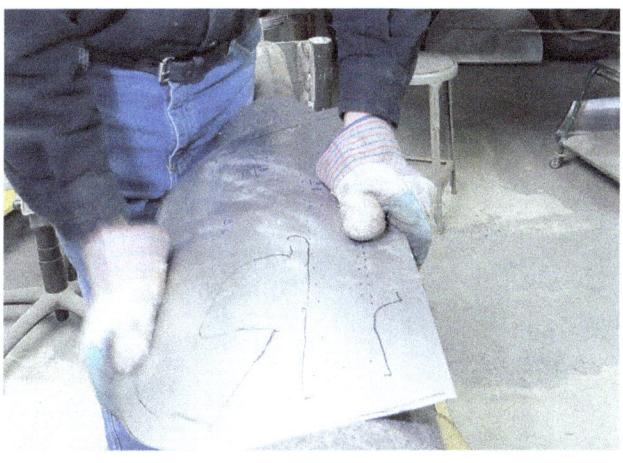

Pressing down over a blanket covered pipe, Wray persuades some curvature into the panel.

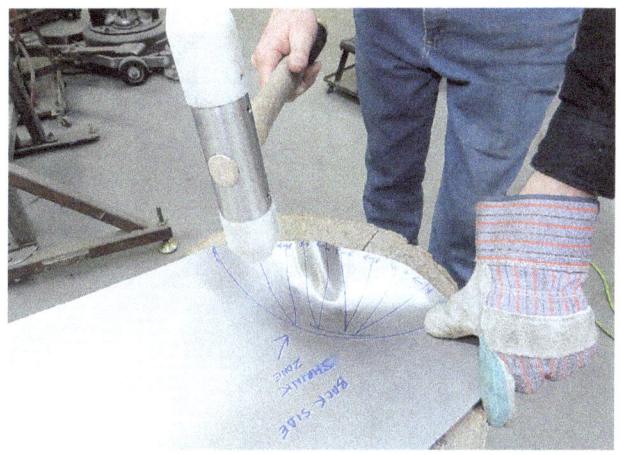

His favorite shrinking method is to use a tree stump. The UHMW hammer is a Schelin design which has been widely, and inaccurately, copied by other manufacturers.

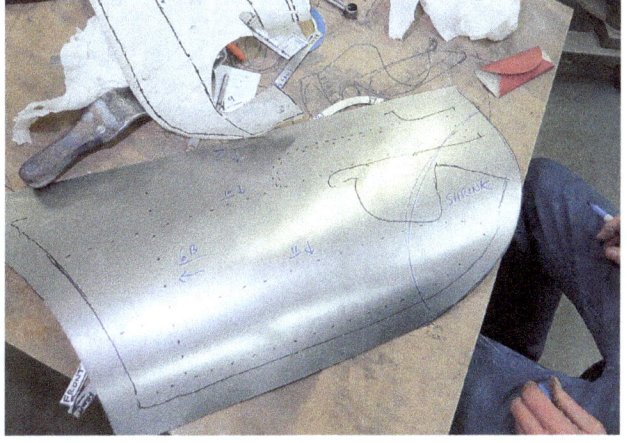

Most of the dotted lines show where the templates go although the "cane" shaped double lines show where a swage must be tipped in. Wray is not sure if he will cut out the opening for the bumper bracket at this stage.

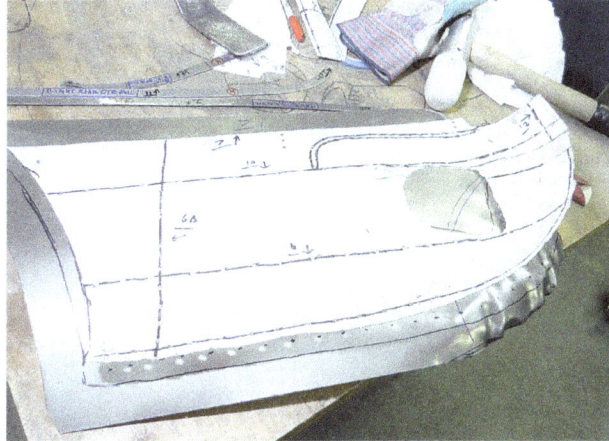

Wray decides not to cut out the bracket opening and adds strips to his FSP. He feels achieving the overall shape at this stage is more important.

The ruffles and walnuts developed on the stump are quickly wheeled out.

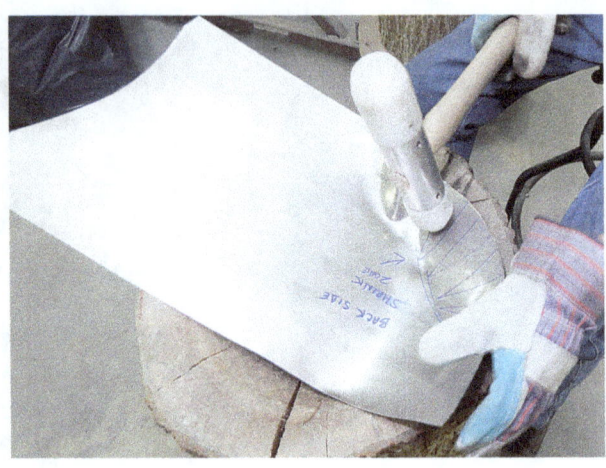
So Wray continues to shrink the edge.

It is easy to get carried away when wheeling, but wheeling too far into the panel would flatten the shape created on the pipe.

This anvil was chosen to get into the tightest corner.

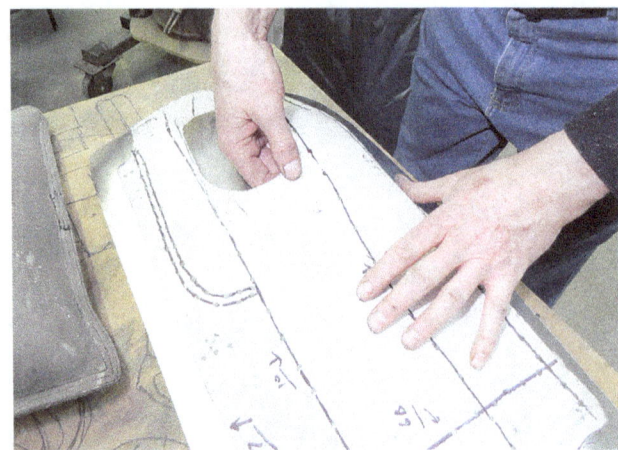
The FSP doesn't perfectly hug the panel yet, especially in front of the bracket opening.

He changes back to the more commonly used flat anvil.

Wray blends the highly curved end of the panel into the center area using just above washover pressure.

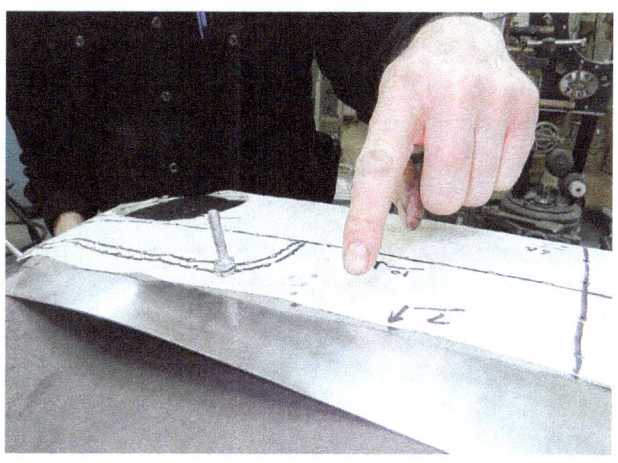
Though the panel looks right, the FSP tells another story. See the loose area under Wray's finger? The panel will have to be raised here.

Holding the panel up to a fluorescent light is a quick way to locate highs and lows.

Which is quickly done by some edge stretching in the wheel.

Some very light wheeling raises a low spot.

Wray adjusts his tipping wheels in preparation for adding the swaged detail.

Wray Schelin: Renaissance Man

In many disciplines you will find people who somehow just see things way before anyone else, and lead the way. Think Benjamin Franklin, Louis Pasteur, or Nicholas Tesla. Wray Schelin is just such a pioneer in the field of panel beating. Wray not only has many machine and process innovations to his credit, but he also was the seminal figure in the explosion of metalworking communities on the world wide web. Literally no one played a more important role in bringing together thousands of panel beaters, and would-be panel beaters from across the globe than Wray. In so doing, he is one of the handful of men most responsible for the metalshaping renaissance that has occurred in the United States. John Glover, Ron Fournier and Kent White are others.

Wray set out to learn metal shaping after learning high end auto restoration as a teenager in the 1960s from his step-grandfather the great Ted Billings, noted Duesenberg expert. In those pre-internet, pre-VHS tape days it was extremely difficult to learn the secrets of the trade. Ron Fournier's book wouldn't appear until 1982. Wray clawed his way to competency, became one of the country's Jag XK body repair gurus, and in so doing developed the power to understand metal in ways that most of us who haven't had to fight for that knowledge ever will.

Wray's development of gargantuan wheeling machines, hyper-adjustable air planishers, silky smooth flanging machines, and myriad other great and small innovations, such as the safe and smooth shrinking disc, has rendered him an honored figure amongst metal cognoscenti. The Schelin Flexible Shape Pattern, which he developed to aid two sculptors with whom he was collaborating on a project, ought to earn him some sort of Panel Beater's Medal of Honor. It was Wray's detailed suggestion that prompted the late Terry Cowan to register the first internet metal community from which sprang others. Wray then went on to suggest the establishment of "metal meets"… a voluntary gathering of anyone wanting to learn metal craft and those willing to teach it for free. Metal meets now occur all over the world nearly every month and these account for the surprising fact that there are more competent metal workers in the world today than a generation ago. Wray is directly responsible for that renaissance.

Wray Schelin, the unsung hero of the panel beater's craft.

The double line of tape shows where the tipping wheel must track to create the swage.

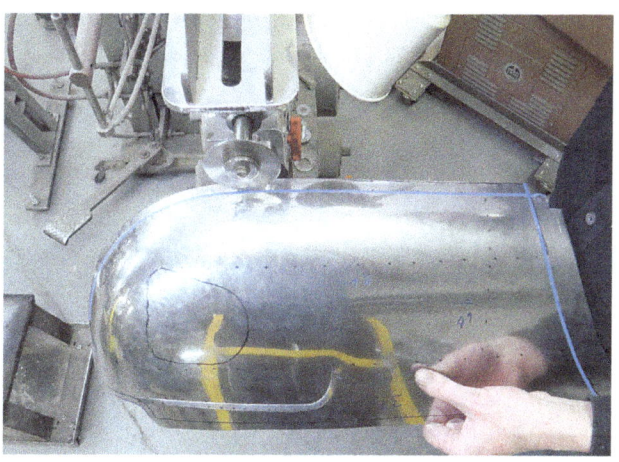
The same technique is now used to turn the flanges.

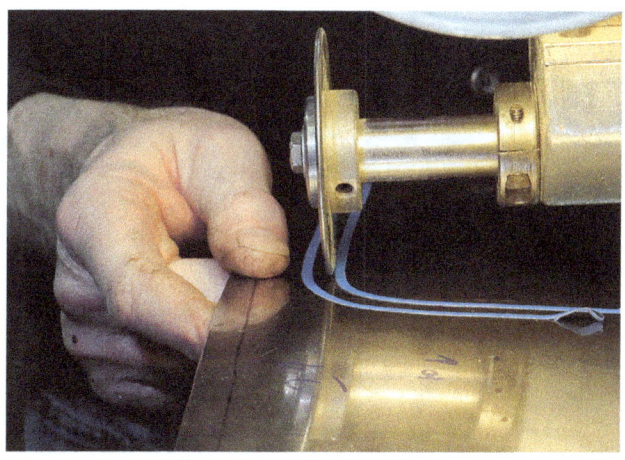
Wray prefers to hand crank his tipping machine to keep perfect control.

As you crank the machine, lift up about 30 degrees per pass.

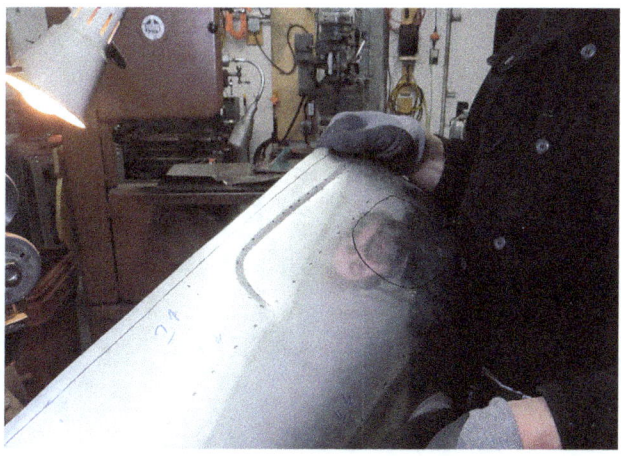
A beautiful swage in exactly the right spot with no drama.

When flanging around a corner, you must shrink the excess metal after each pass.

Wray makes the flange line crisper with a blunted chisel and slapper.

A tricky double-flange is created with the thin tipping die.

Using a different planisher, he smoothes out any remaining irregularities in the panel's surface. Wray's attention to detail is unexcelled.

Drilling out the hole for the bumper bracket is easy with Irvin step bits. Wray avoids the cheap imported step bits. A nibbler expands the hole to its final shape.

A close-up of the anvil he designed so as to get into the tightest flange corners.

Not welded yet, just checking the symmetry.

This part of the original panel will be saved. The FSP was created specifically to replicate the missing side.

Wipe your wheels if you want to avoid scratching the panel.

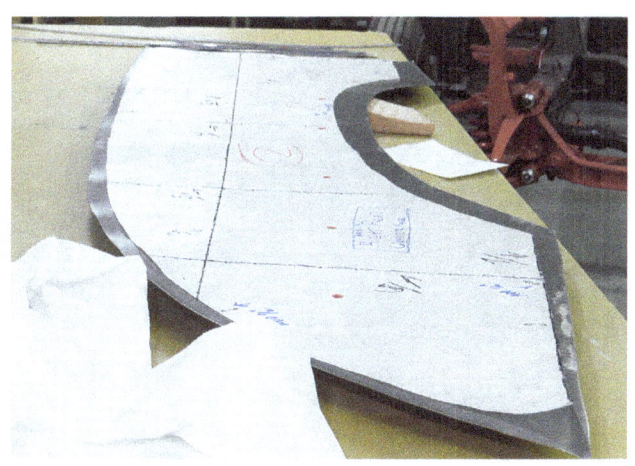

Waviness along the curvature of the FSP shows where shape is needed.

Wheeling begins on the edge so as to stretch enough material to round over.

Wray uses a straightedge to verify the panel's tucking in at the rear.

This five foot panel would be best shaped by a two person team, but Wray is a big guy with a large arm span. That huge upper wheel also helps keep it under control.

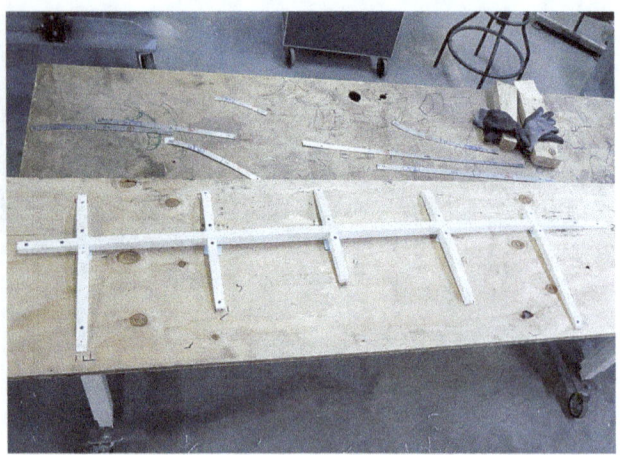

This compound tapered support for the panel is not really a buck as it won't be used to shape the panel. Rather it is a support for the otherwise floppy panel when using the FSP to make shaping decisions.

Wray positions himself a step back from the center of the wheels so that he can swing the panel through on a curve. He keeps his hands up and level during the movement to avoid wheeling a twist in the panel.

A panel this long, with such subtle shape, would flop mostly flat on the table without the support skeleton, thus rendering the FSP difficult to read.

*The waviness of the panel's upper edge is necessary here. By stretching the edge, Wray has increased the panel's **area**. When he rolls over the excess to form the curved edge, he will have changed the panel's **arrangement**.*

Achieving such a gentle curve over the length of the panel requires a lot of careful wheeling. The template is vital to attaining a perfect panel.

Wray preaches the concept of Area and Arrangement in his metal classes. He also practices what he preaches.

The template indicates more raising is required.

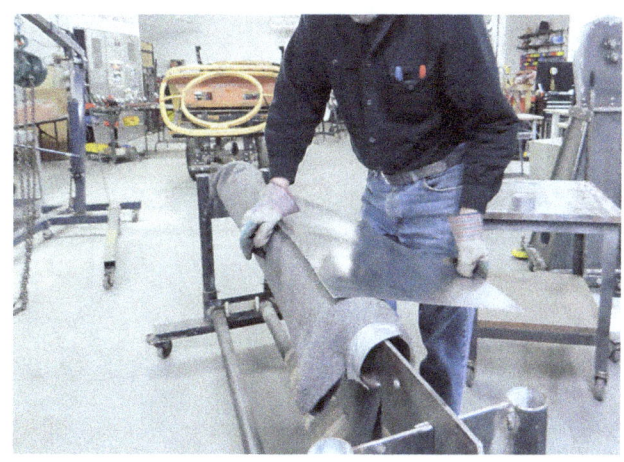

Back on the pipe, he continues to force the edge over.

So he takes it back to the wheel and does some short cross tracking a few inches in from the upper edge. Through experience he judges the upper edge has sufficient stretch already to go over.

He blends the shape created on the pipe into the rest of the panel. He is working very slowly, and thoughtfully here.

With a flat wheel, and minimal pressure, he inserts the upper edge into the wheels and gently pulls down and out. This begins to curve over the edge.

At this point, Wray has several hours in the panel. His greatest concern is not to over-shape it through too much wheeling.

The upper rear of the panel needs to come over more, so he stretches it to create more area.

Hand manipulation is part of the process - most involve minimizing twist. Here he tightens the upper curvature. RIGHT. Wheeling in more shape as previously indicated by the template.

Now that the upper curvature has been formed along the entire length, the panel is much stiffer and easier to wheel.

Finally satisfied with the shape, he lays out flange lines with tape.

The left template is spot on, but the right shows the panel needs more body near Wray's thumb and along the upper edge.

After turning the flange, he places a strip of copper under the join lines in preparation for welding.

Wray prefers to TIG his panels and says he's had no cracking issues over the years.

...the Sharpie ink lets him know where he's heated the panel and acts as a mild lubricant. Once the panel is heated, he quenches it with a water mist. The results are impressive.

One of Wray's most useful innovations was the safe and smooth shrinking disc. Here he uses a wide tip Sharpie to mark the entire area to be shrunk with a disc.

The weld is then dressed with slapper and dolly.

Wray improved the utility and safety of the old bodyman's shrinking disc by removing the waves, and rolling the edge. Heating the panel with the disc not only tightens loose areas, it stress relieves the panel...

Where once there was nothing, a new quarter panel emerges.

Chapter Ten

Formula II Cooper Repair

A New Nose from an old Buck

John Cooper's little car company in Surbiton, England forever changed motor racing with their almost accidental introduction of rear-engined racers in the 1950s. Peter Tommasini was tasked with building a new body for one of these little giant killers, and makes the process look very simple.

Pay particular attention to how Peter uses nothing more than his e-wheel to precisely locate the detail line that separates the top of the car from the side. After creating the line, he stretches the flange to create the reverse. The length of the Cooper panel is really at the limits for one man.

Where are the wings, the corporate logos, the on-board telemetry? A lovely Cooper racer from back in the days when John Cooper said, "racing was fun."

This buck has seen prior service to this build, and will be used again. Such is the fate of race car bucks.

Understand this: No shaping has yet been done, only folding, and yet the basic form is apparent. The four things you can do to metal are: cut, fold, stretch, or shrink. The latter two are "shaping".

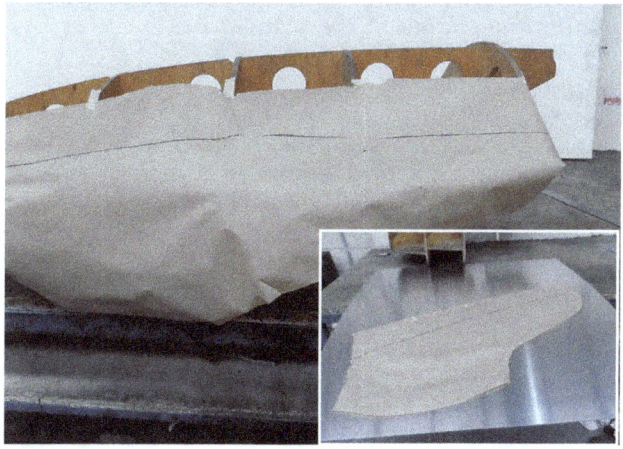

Even on a highly bulbous shape a paper pattern is a very accurate tool with which to lay out the aluminum blank. INSET: The line on the pattern defines the longitudinal apex.

By striking over the hollowed out "bowl" of the stump, he raises ruffles which will be shrunk to tighten the edge. These medieval armorer's techniques were used to make the great race cars of the 20th century.

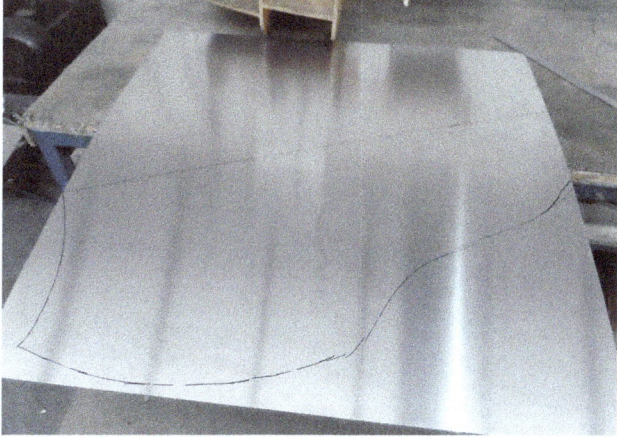

Many race cars were made from .050 aluminum, but ultra-competitive Enzo Ferrari was known to specify .020 sheet!

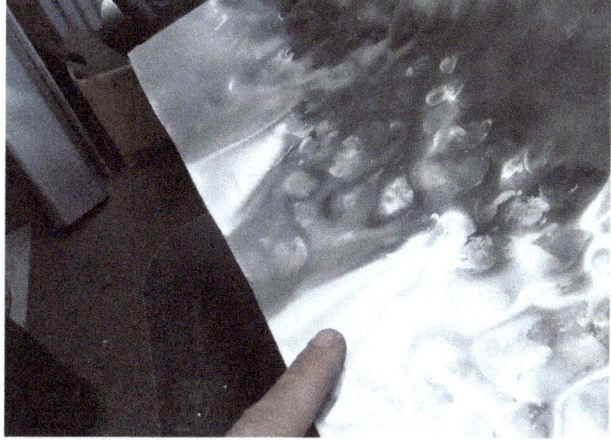

Peter points to a "V" shaped ruffle. You want to create narrow, more defined ruffles than the rounder, more open ones in the previous photo.

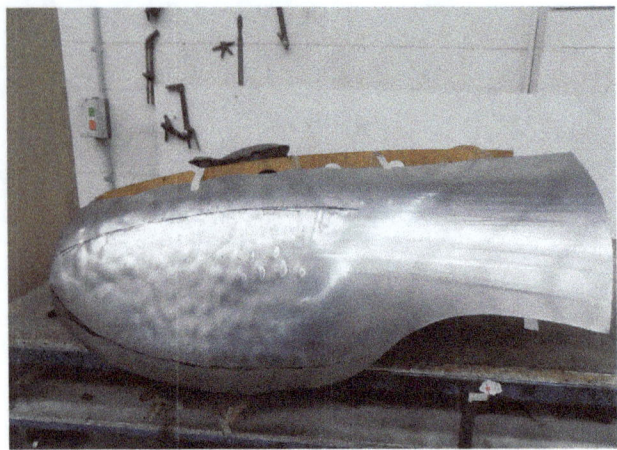

Some wheeling has been done on the right of the panel to smooth out the initial sharp fold, but most of the shaping so far is in the "bag of walnuts" on the left.

With the shape roughed in he blends the 2 major zones with light wheeling. He uses more pressure to further raise the forward bulge in the center. RIGHT: Looking at the curvature you might think he's using a highly radiused anvil, but it's the near flat #2.

Wheeling out the walnuts along the apex. Peter pulls down somewhat as he pulls out thus increasing the shape.

Wheeling fore and aft here tightens the longitudinal curvature.

On the shop floor, he pushes down on the edge to open the panel a bit. The wheeling on the left has been mostly vertical while on the right longitudinal.

The wheel not only helps turn the panel under, it planishes at the same time.

In the golden age of racing, 1930-60, car bodies were rushed to completion as owners knew they would be pranged very quickly. These days, owners expect the sort of polish Peter is developing through close tracking.

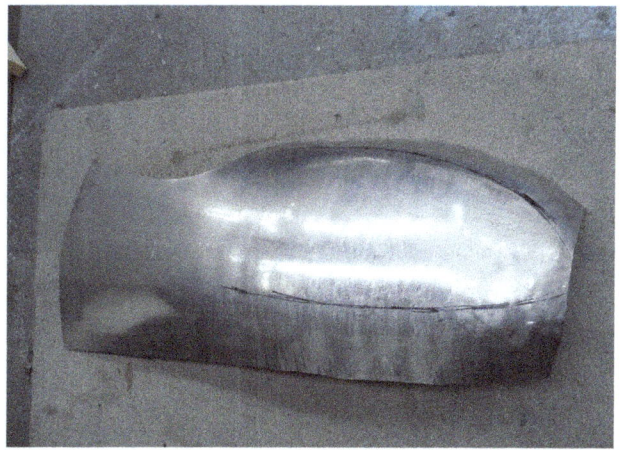

The area (shape) is 95% in the panel now, but the arrangement is not.

He switches to the No. 4 wheel to crank more shape into the lower edge of the panel where it tucks under the body.

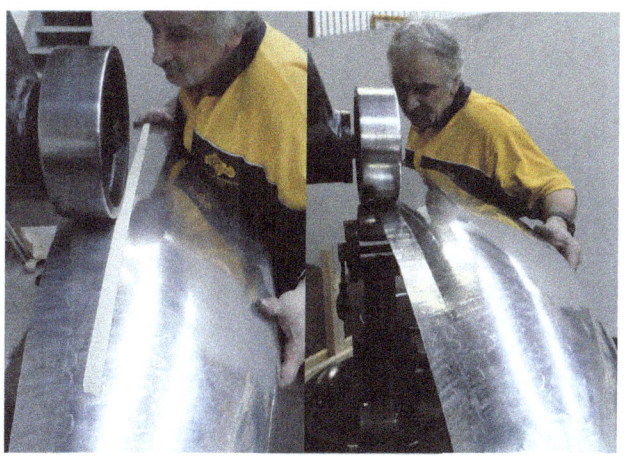

Peter follows a tape line to create the reverse needed to meet the upper panel. RIGHT: Flattening of the panel here is a change of arrangement, - the panel hasn't changed its total amount of area, only that area's form.

A cross-wheeled pass sucks in both edges of the panel, and relieves developed stresses.

However, wheeling along the flattened flange now with a flat wheel and medium pressure, he adds area (stretches) the flange, and gives it enough area to turn over to join the upper panel.

Some waviness has crept into the panel when the flange was stretched. Peter levels this with some close cross-wheeling.

The nose is tight, but needed some mechanical shrinking.

After checking against the buck, he wheels a little more shape into the highest point of the crown.

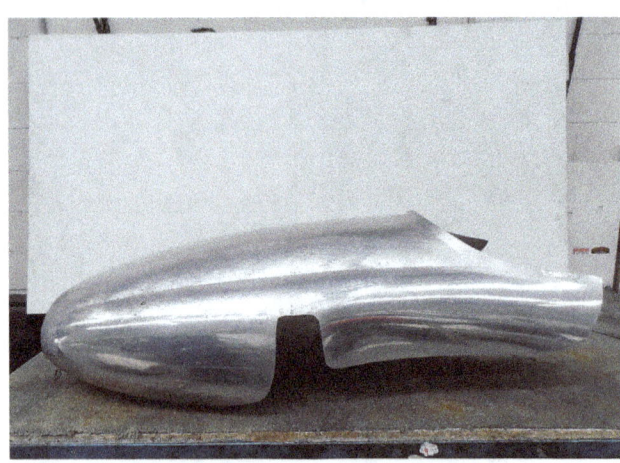

The right side was made similarly to the left, but the upper panel was easy to form because it was all crown except for the small reverse at the cockpit.

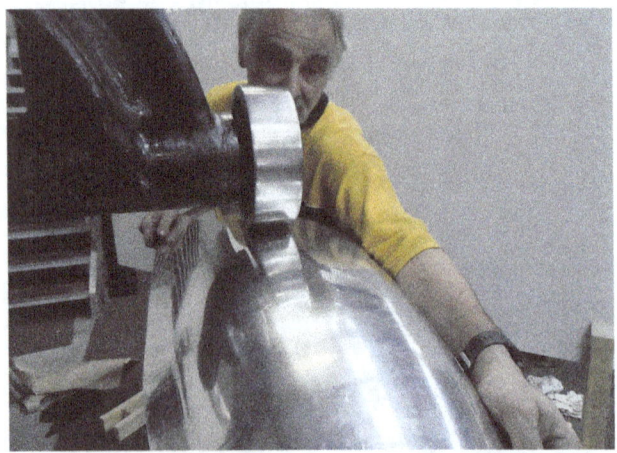

Final close tracking develops a mirror image.

The new nose is ready to go racing and was built using techniques John Cooper would have used in his Surbiton shop.

Peter Tommasini: Ciao Maglio, G'day E-Wheel

Pierluigi Tommasini hailed from Mestre, Italy, a town near Venice, and began his metal apprenticeship at age fifteen with none other than Scaglietti, the carrozzeria much favored by Enzo Ferrari during the company's glory years. In those days of noisy, welding-fume-filled, summer sweltering factories where bullying the apprentice was commonplace, and cacophonous maglios dominated the shop floor, the gig at Scaglietti was not as romantic and as it might seem in rose-colored hindsight. Pierluigi's family soon immigrated to Australia and took him with them.

In addition to learning a new language, Pierluigi, now Peter, had to learn a new style of panel beating. The Anvil Chorus pounding of the maglios gave way to the Elgar symphony of English wheels. With a friendly, easy going personality Peter adapted well to the change, but found, like many others before and since, that mastering metal was not something he could expect to be taught, but rather something he had to learn.

While doing mundane collision repair for several years during the day, Peter applied himself to mastering real metalshaping at night. He was greatly aided by Tom Peach, Bob Head, and Gary Tishler, Aussie metal masters raised in the English panel beating tradition. By age 30 Peter had learned enough to open up his own shop specializing in panel fabrication for classic cars. In the three decades since then he's done them all from Austro-Daimlers to Zagato bodied Ferraris.

The Australian TV network Channel 31 reunited Peter with Tom Peach to create a popular program called Gasolene in which the pair taught traditional panel skills. The show brought Peter well-deserved national prominence and he toured the country giving seminars on the subject. Eventually he offered lessons at his shop's home in Melbourne, as well as a line of instructional DVDs and tools. His DVD Series 7, 8 & 9, during which he builds an entire Monaro quarter panel from a single piece of steel, without welding, will blow you out of your seat and cause you to think that maybe it's time for you to emigrate to Oz.

Peter began his trade in Italy where shops used maglios like this one. In Australia, he learned to use English wheels, which he much prefers.

117

Chapter Eleven

Randy Ferguson

Where There's A Willys, There's a Wray

Though Randy Ferguson had a thriving trade as a respected collision repairman and car painter, he wanted to move beyond Bondo mud and paint fumes. One day, back in 2000, while fooling around on that new-fangled invention "the internet" in his home in Robinson, Illinois, he did a search for the term "metal shaping" and up popped the name Wray Schelin. In a remarkable coincidence, Terry Cowan, on literally the same day, had set up the first internet metal shaping group at the behest of Wray. And serendipitously Randy found it.

Randy specializes in turning rusted out Willys hulks from this...

Randy entered a voluminous e-mail correspondence with Wray, and other members of the nascent group, learned as much as one could about the theory of metalworking through cyberspace, and like the wanderer who was lost in the desert for years, knew he had finally found home. Only those who craved to learn panel beating before the advent of the www can understand Randy's sense of joy, and relief, at this point in his story.

In the following months Randy soaked up metalworking knowledge and began to practice the new techniques he was learning. He finally met Wray at the forerunner to all metal meets called FormFest in 2001, in Huntsville, Alabama and the two became fast friends. By 2003 Randy's skillset was already impressive and he began hosting MetalMeets himself at his shop in Illinois. During all this he made the big decision to leave collision repair and establish a full-time business building panels for classics. He evolved a specialty for Willys coupes and within a few years he became the go-to guy for any panel for that car in the country.

Randy is effusive in his praise for Wray's tutelage, but I can attest that Randy himself has become a great teacher. Thanks to Randy, Mark Stewart's remarkable Daytona Coupe build became an Oblong Metal Meet group project, and numerous would-be wheelers were able to hone their skills and move up a level because of it.

Randy is one of those guys who, despite living in a town so rural it's on the fringe of cell phone reception, and having had no local mentors to turn to, had the will to learn the craft, and so overcame all obstacles to achieve the success he enjoys today.

There is a cult following for pre-war Willys coupes driven by those who appreciate the elegant simplicity of the body lines, and the car's light weight which makes if perfect for racing. Willys really are something special, whether restored stock, or as often is the case, turned into hotrods. Since they were made in such small numbers compared to Fords and Chevys, replacement body panels are nearly impossible to find, so Randy is often called on to make new ones from scratch.

...into this.

As a disciple of Wray Schelin most of his panels are laid out using an FSP.

He carefully deburrs the panel using a Dan Shady designed tool.

Some longitudinal passes through his monster-sized e-wheel quickly begin to raise shape in the panel.

Randy does his first test fitting with the FSP and finds the center of the panel needs raising. This FSP serves both left and right side rear fenders and has been used many times.

As he rolls, he mentally visualizes the "no-blow-zone" around the perimeter and avoids it.

This close up shows the bubbly condition of the FSP which indicates the panel is too low.

In just a few minutes the panel is stiff enough to not flop over the No.2 anvil and has vertical and horizontal curvature.

Randy follows the curvature of the panel through the wheels, but still avoids the edges.

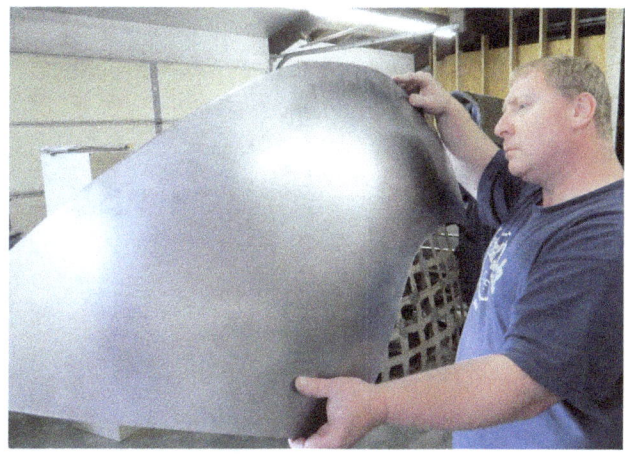

The crown is coming up fast now, but eyes alone are an unreliable judge of shape.

Still using the No. 2 anvil, what he calls his "go-to" anvil, he wheels tangentially to the panel's curve, thus bringing the upper and lower edges towards each other.

The FSP is much more settled now, but the upper edge has become wavy. The cure is to wheel (stretch) just inboard of the waves, and this is verified by the bubbles in the FSP there.

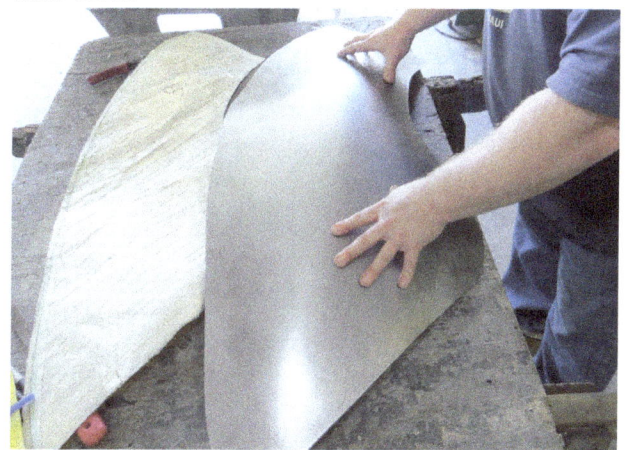

After verifying with the FSP he deems the panel finished. Total time: 40 minutes.

Satisfied that he's close now, Randy trims off excess material.

He lays out the next blank, the lower front of the fender, using an FSP.

Tangential wheeling along the curve helps create the tighter shape of this panel more quickly.

You can tell by the dullish ring in the center of the upper wheel that Randy is using flat cut anvils. Most, but not all, e-wheelers prefer these.

You can see that he has avoided wheeling the edges, and now concentrates on raising the center.

The dip in the panel next to his forearm is due to too much area (stretch) there. If you have a wavy edge, wheel inboard of it. (See the Shape In-Shape Out exercise.)

The first test fit with the FSP tells him he's way too low in the lower center. Solution: wheel more there.

Hand-Knee manipulation removes a slight twist.

The FSP fits well, so the next step is to mark the panel and trim it.

On a hollowed stump he creates edge ruffles which he'll shrink quickly to create a curvature in the panel.

The steel blank for the upper front is prepared by cutting and wiping off the oil.

The ruffles are shrunk from inboard towards the edges with each blow taking in about 3/8 inch (10mm). Though Randy uses the stump here, steel forming heads are better.

Randy does some quick wheeling at a 45 degree angle because wheeling one way the tracks are too short, and the other they're too long.

With the initial shaping done on the stump, he returns to the wheel to planish before shaping again.

He finds he needs even more edge shrink, so it's back to the stump.

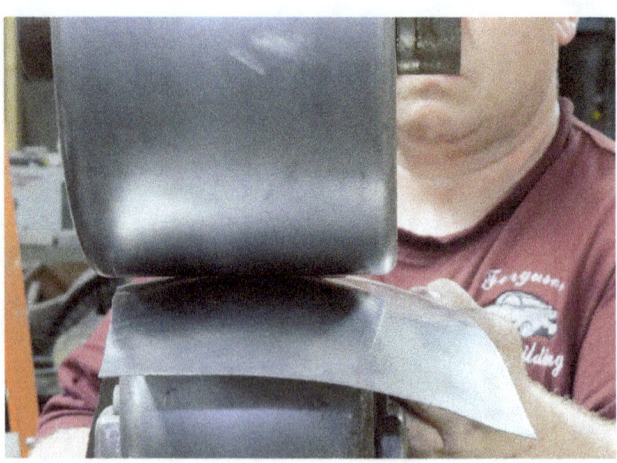

Finally he begins to stretch the center of the panel by wheeling.

All of the shape you see in this photo comes from shrinking on the stump. The steel molecules along the edges are literally packed tighter through this process.

Your hand must help guide the panel through the wheel respecting the shape of it. Here, Randy's hands must go up and down as the panel goes in and out.

After planishing in the wheel, the ghosts of the ruffles are barely visible, but cannot be felt.

Careful here! It is so easy to let the panel slip into the wheels right up to its edge. If he allowed that, he'd lose all of the shrink he gathered on the stump.

Because the panel is fairly close to its final shape, he shrinks the edges a bit more using a Lancaster shrinker.

The rear panel is clamped in position and will be trimmed to fit the front.

The part fits nicely, and only requires some trimming.

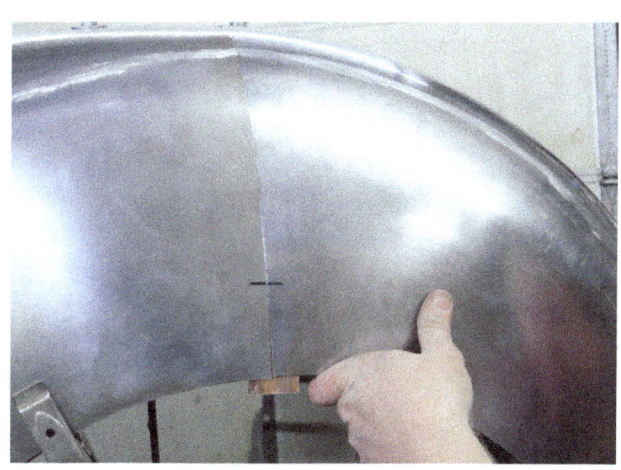

After trimming, he inserts a copper strip behind the seam to protect the Bondo from rapid off-gassing and popping. Note the alignment hash mark.

The two front pieces are clamped to a Bondo buck.

Randy prefers TIG welding his panels, and does so without an elbow rest.

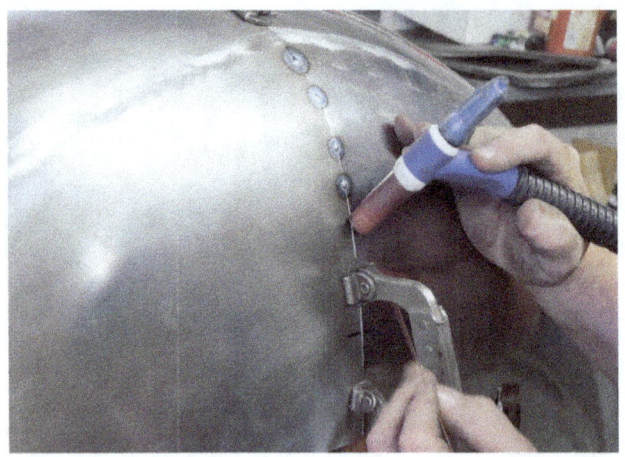

He first lays tacks about every ¾ inch.

Whenever possible he prefers to weld on a horizontal plane.

Before finish welding he grinds off the heads of the tacks.

Any weldment that is proud of the panel gets cut down with the grinding wheel.

Next, he confirms absolute panel alignment with a body hammer. RIGHT: Do you think that big wheel has any trouble crushing this weld seam?

The weld seams are given a light initial sanding with 120 grit.

Randy Ferguson

Probably some of the best advice I received as a novice sheet metal shaper came from my good friend and mentor, Wray Schelin. That advice was twofold, really. The first part was to start studying shape. For the most part, everything we see has shape and/or form. In the realm of sheet metal shaping, we view shape as a change in surface area, producing a smooth, appealing object, whereas form is something as simple as bend, fold, simple radius, etc, in other words, there is zero to very minimal change in the overall surface area. By studying shape, you train yourself to analyze an object toward making a decision as to how you would copy that shape in metal. Learning to understand how an object flows and what goes into reproducing that shape gives the shaper insight as to how to approach any project.

The second part of this is that you do not need a shop full of costly equipment to produce even the highest quality products. A few hand tools are all that's necessary to get you started. As you gain experience, then you can start adding more tools to make the process move along a little quicker.

Of course the flip-side to all of this is that you must invest the time to learn how to become proficient at shaping sheet metal. Sitting around thinking about it, reading about it, watching videos about it will only give you certain amount of intellect on the subject. Not that that is a bad thing, but all the intellect in the world will never allow you to put it to practice without the experience to go along with it.

Beyond this, the best advice, perhaps, I may be able to offer is to find a good mentor. One who will give you good advice based on his/her own experiences and will guide you through the process at your pace and with the understanding that you may only have a few hand tools is the one you're looking for.

Randy Ferguson saves another Willys.

Chapter Eleven - Part 2

An Important Panel for all Metalshapers

An Anticlastic Panel

The Willys coupe was a continuation of a design theme pioneered in the U.S. called the "pontoon" style. Cars of this design disposed of running boards and open fenders in favor of a more streamlined body with wheel-encasing fenders. The "coffin nosed" Cord is considered by many to be the pioneer in this design.

The style began to spread to Europe before the war, most notably with the hugely influential BMW 328, but after the war an even more all-enveloping style evolved called pontonkarrosserie. Bulbous fender fronts and grilles flowed back into slab-sided bodies. The Italian design houses, especially Farina and Vignale, perfected the theme.

A feature common to all of these cars is the anticlastic panel between the fender and the grille opening. In other words, a saddle panel. Forming this shape must be mastered by any would-be panel beater who hopes to consider himself a professional. Randy shows us how to do it.

Designs like this pre-war BMW 328 helped spread the "pontoon" design theme which first became popular in the U.S.

The Farina designed Cisitalia 202 is considered by many to be the most influential post-war car design. It featured "ponton" styling and a very subtle anti-clastic front panel.

Using his e-wheel with the soft upper wheel and hard plastic anvil he forms a gentle curve in the panel. Done in a normal e-wheel would cause the panel to curve in longitudinally downwards...

Randy holds a well-worn FSP up to the Willy's "saddle" area between the grille and headlight.

...not just laterally, and therefore opposite of what he needs. He moves the panel side to side to spread the curvature across the panel.

He cuts a blank from steel.

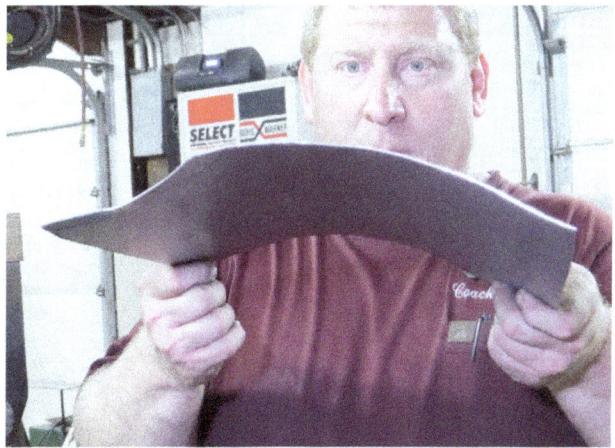

This shape could have been created on a slip roller, or even over a welding tank. Randy's e-wheel made the process predictable and successful on the first attempt.

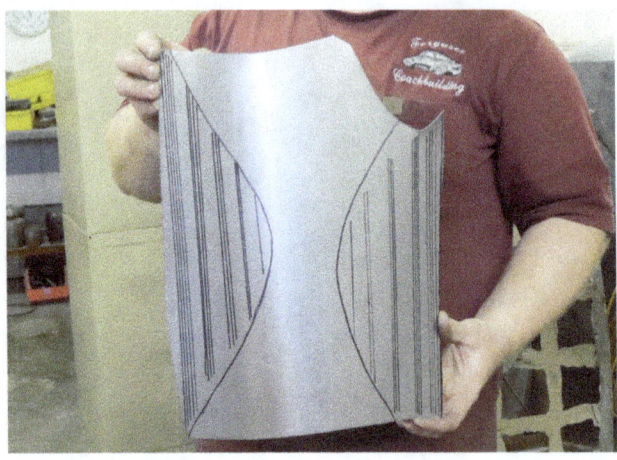

These drawn lines show where the majority of the wheeling will occur. On all reverses the goal is to stretch the edges with more passes than the center.

You've got to pull up on the ends to get the panel to move where you want it.

Notice the panel is being wheeled from the back side now with medium pressure.

Even though the lines suggest that the wheeling remains inside the drawn curve, you've got to do a little blending into the outer area to avoid humping the panel.

With only a few minutes work the panel starts to dip downwards as more area is created closer to the edge (stretching).

Lifting the front…

...and lifting the back. Randy is using medium to high wheel pressure here.

Imagine the drawn pattern from the back now drawn on the front. That is how the panel will now be wheeled.

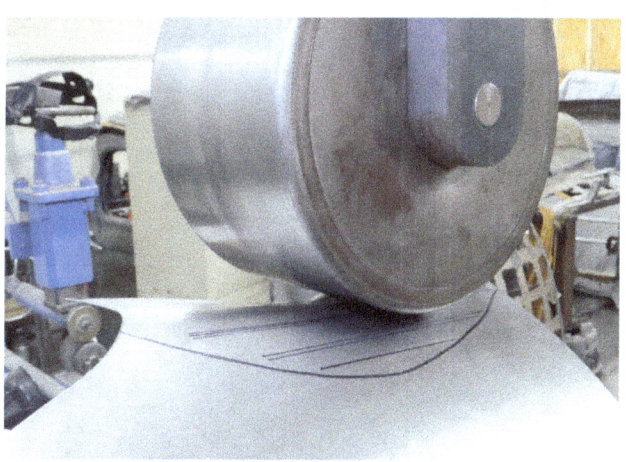
A good practice is to develop both sides of the saddle evenly.

By rotating the panel 90 degrees and wheeling on the front side, the panel is brought even further around through edge stretching.

An anticlastic shape cannot hold water no matter how it is positioned.

Finesse this edge so as not to over-curve the headlight mound.

The FSP is close, but some looseness is evident near the top.

So Randy creates more area by wheeling the edge.

The panel is flipped again for further "Chinese wheeling."

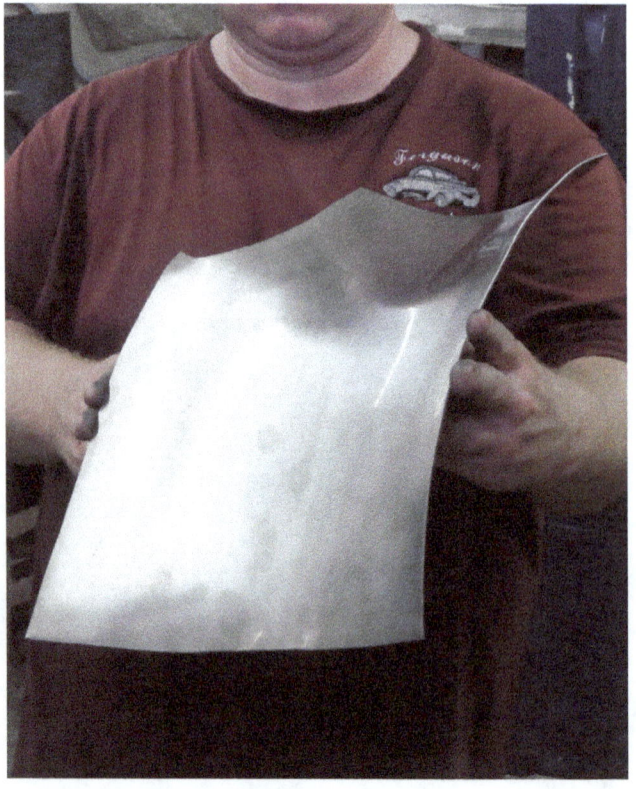

If you can't do this you can't build most cars from 1940-1960.

Along with the '32 Ford, the Willys coupe is considered one of the "holy grails" of hot rodding. This one will live again thanks to Randy.

Excellent fit and all that remains is turning flanges.

Chapter Twelve

Kent White

Wings, Wheels & Welding

If you think wheeling panels for billionaire owners of Pebble Beach Concours contenders is demanding, try wheeling panels for aircraft… real aircraft that fly and whose pilots and passengers' lives depend on your work. Suddenly the art becomes a science. Those panels need to be wheeled and tempered to certifiable standards, and such certified work can only be done by a very few, the crème de la crème. Who ya gonna call? Kent White, aka The Tin Man.

The metalworking renaissance in America that I've spoken of earlier owes much to Kent who is

Aircraft Art in Aluminum. Jim Wright's magnificent, yet fated, Hughes H-1 replica. Wright called on Kent to wheel the most difficult panels.

responsible for recording on VHS tape in the 1990s a significant syllabus of metalworking lessons that did far more to spread the gospel than mere words and pictures found in books. Kent filled his video how-to lessons with logical step-by-step sequences, dramatic challenges… such as the time he took a hatchet to a Porsche RSK Spyder nose-and then repaired it, and a barrel full of Mark Twainish humor. The real power of Kent's videos is that though the viewer feels he/she is being entertained, they are really getting a schooling in both the art, and science, of metal.

Kent began his career in the fabulous workshop of Bill Harrah, the Nevada casino tycoon who amassed a legendary car and aircraft collection in the 1950s-70s. He hired the best "old-school" metalmen from the pre-war years to work in his spare-no-expense restoration facility, and it was from these that young Kent began to learn his trade. Kent's own thirst for knowledge led him to go on to study the scientific underpinnings of metallurgy, welding, and structures, and it is partially with the eye of a craftsman, and partially with the mind of an engineer, that he approaches metalwork. The reputation he built over the years has led him to become a Boeing certified repair instructor, an in-demand panel beater for historic aircraft restorations, and a consultant to metal shops of major airplane companies.

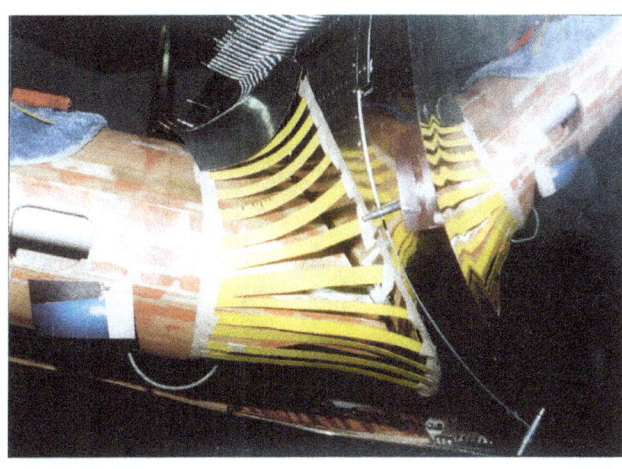

Kent's solution for laying out the reverse-shaped panel that would be the leading edge wing root fairing.

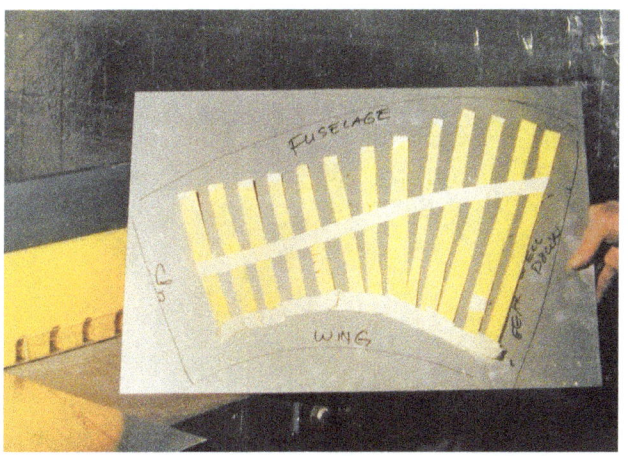

Strips of poster board held together by masking tape. Leave a border for final trimming. Move one tape strip inboard to transfer finger pattern onto blank.

All four sections adjoining the edges will have to be stretched to create a "saddle" or "reverse." The top and bottom sections will stretch rear-ward while the left and right sections come forward.

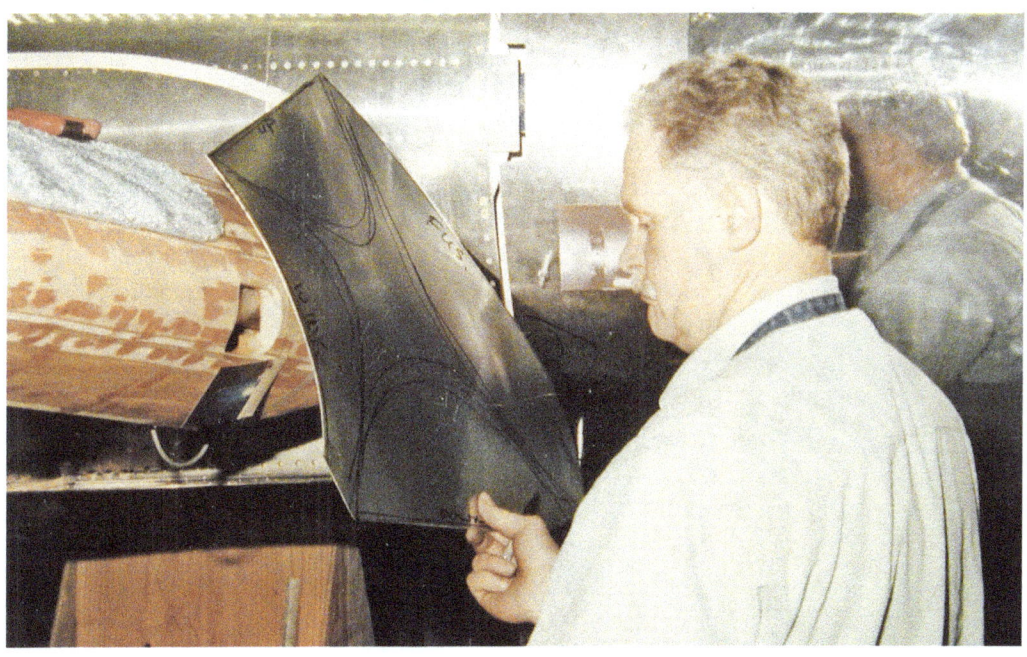

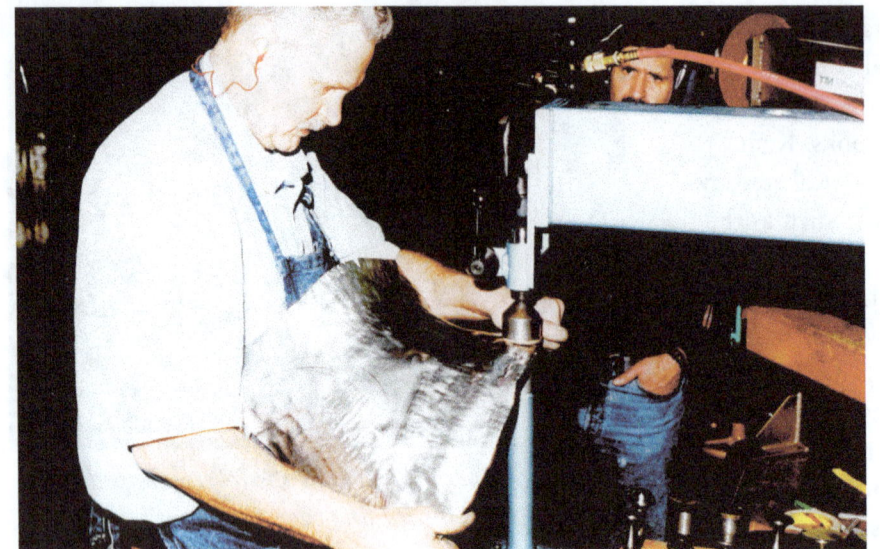

Kent uses linear stretching dies in one of his highly regarded air planishers to rough the panel into shape. On "anticlastic" reverse panels, you concentrate the stretch from the edges inward, tapering to zero.

Kent uses one of his T-M Technologies e-wheels to planish the stretch marks and blend the four contoured zones to fit perfectly.

Kent's e-wheels are used in airline repair shops around the world. Though capable of heavy duty work, they are also portable.

The leading edge fairing is held in place by one finger. It's a perfect fit between the aluminum fuselage and wooden wing.

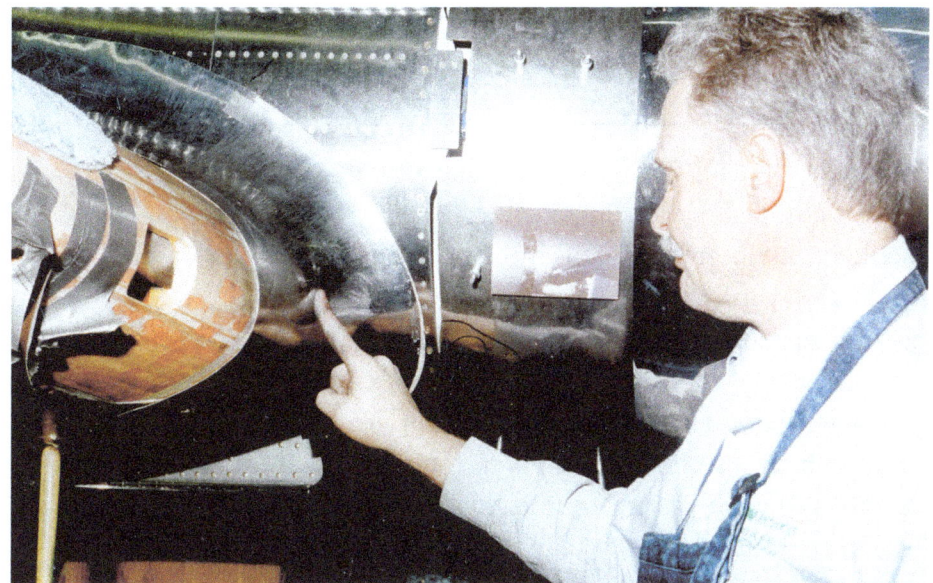

Wing fairings reduce the air turbulence between the fuselage and wing. This airplane was capable of 360mph in flight, and so the panel's fit and contour was critical to limit parasitic drag.

When people's lives depend on your welding skills, you'd better be this good. Kent gas welded this seam on a 1937 Waco.

White's Welding Wisdom

Before we end this book, let's take a brief look at aluminum welding, a subject we'll tackle in much greater detail in my next book.

You've seen that the craftsmen I've showcased so far differ on their welding preferences, some go with gas, and others with TIG. Kent has more than just an opinion on the choice because he has studied the topic to the point of consulting the craftsmen and the technical literature on it. He has to take the subject seriously because often his panels wind up on aircraft which fly, and therefore people's lives are dependent on his welds.

His research, both empirical and scholarly, has proven to him that gas welding is the best method to use with the aluminum work he does. He has performed destructive tests on both TIG and gas welds, and the TIG seams always fail before the gas ones. The research published by the Aluminum Association in

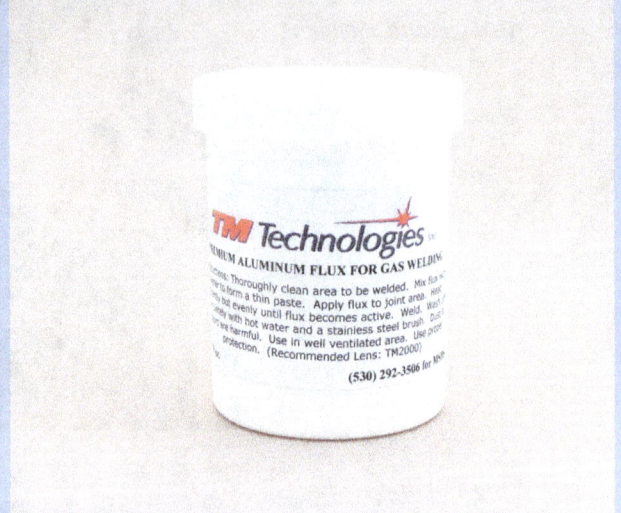

You can't gas weld aluminum without the proper flux. Get it from Kent at www.tinmantech.com . A little goes a long way.

My preferred gas welding setup is a Purox W-200 with a Kent White lightweight hose. Mark Barton and Steve Hall also use this torch.

How much are your eyes worth? I strongly urge you to get his TM-2000 lens rather than cobalts.

White's Welding Wisdom

America points out that electric welds are always less dense than gas welds, and that is one of the reasons they are more prone to porosity and leaking than a gas-welded joint. The other reason is the inherent bulkiness of a TIG weld.

Kent strongly recommends using a small, classic-style welding torch, such as a Purox W-200, Smith Airline AW1A, or Meco Midget, in preference to the low pressure torches such as the Dillon/Cobra/Henrob series. The Meco Midget was designed almost 10 years after the Henrob! Again, he cites weld-bead issues with the latter pistol-style. He points out that some major aircraft manufacturers still gas weld, and those who don't use very expensive custom TIG setups not to be found in smaller shops.

From my personal experience I know that the reason many panel beaters use TIG welding is that gas welding takes continued practice to stay proficient. TIG welding is somewhat less demanding. However, in this instance, I have to say that "old school rules" and the best (most workable) aluminum welds will be achieved with gas.

Wolfgang Books On The Web
http://www.wolfpub.com

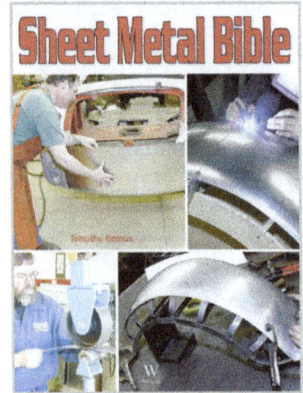

SHEET METAL BIBLE

Sheet Metal Bible is a compendium of sheet metal fabrication projects, everything from simple shaping operations to multi-piece creations like fenders and motorcycle gas tanks. Each of these operations is photographed in detail. Meaty captions help the reader to understand what's really happening as a flat sheet of steel slowly morphs into the convex side of a gas tank.

While some of the craftsmen work with hand tools, others prefer the English Wheel. The book is filled with work by legendary fabricators like Ron Covell, Craig Naff, Rob Roehl and Bruce Terry. Projects include components for two and four-wheeled hot rods. Each metal has its place in the metal shop, and this new book includes tips on how to work with, and weld, both metals.

Ten Chapters   176 Pages $29.95   Over 350 color images - 100% color

SHEET METAL FABRICATION BASICS

Sheet Metal Fabrication Basics is designed to provide how-to help for fabricators working at home and in small shops. Whether the goal is a handmade air dam or a one-off air cleaner cover, the project can be done without expensive power tools. This book was created with the help of fabricators like Rob Roehl, tin-man for bike-builder Donnie Smith, as well as Bruce Terry and Ron Covell. Projects include the creation of simple shapes made with

hammer and dolly, hand-operated shrinker-stretcher, English wheel, and various mallets. Some objects are made from more than one piece, thus the book also includes a section on welding both aluminum and steel, with gas and TIG.

You can learn to fabricate sheet metal. All you need is the raw material, the sincere desire to learn and a copy of this book from Wolfgang Pub. 144 pages long, and over 400 B&W images.

Eleven Chapters   144 Pages $24.95   Over 400 photos, 100% color

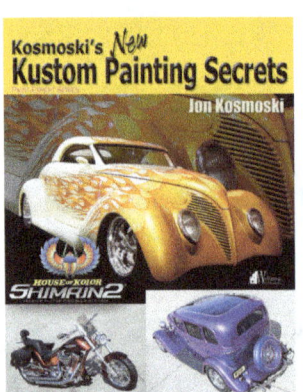

KOSMOSKI'S NEW KUSTOM PAINTING SECRETS

No how-to-paint would be complete without at least two, start-to-finish paint jobs. Kosmoski's New Kustom Painting Secrets contains both a hot rod and a motorcycle paint job. The sequences start at the very beginning with the metal preparation, and moves through all the primer and filler stages necessary to make the panels perfect, before any

topcoat can be applied. The paint jobs include artwork and flames, application of the candy paint, and the final clearcoats. Kosmoski's New Kustom Painting Secrets uses over 400 color images and 144 pages to explain a lifetime's worth of custom painting experience.

Eight Chapters   144 Pages $27.95   Over 400 photos, 100% color

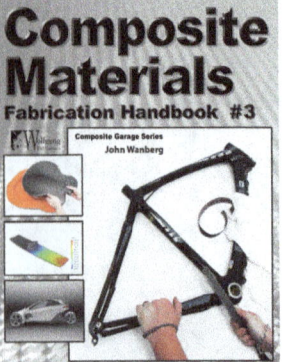

COMPOSITE MATERIALS FABRICATION HANDBOOK #3

Composite Fabrication Handbook #3 continues this practical, hands-on series on composites with helpful how-to projects that cover a variety of topics geared toward assisting home-builders in completing their composite projects. Handbook #3 starts off where Handbook #2 ended, expanding on mold-making techniques including special methods for creating molds and composite copies of existing

parts, fabricating molds from clay models, and making advanced mold systems using computer modeling software. Several alternative methods of fabricating one-off parts are presented in this book, including molding over frameworks and human forms, as well as using stock composites to build simple structures.

Nine Chapters   144 Pages $27.95   Over 500 photos, 100% color

Wolfgang Publication Titles
For a current list visit our website at www.wolfpub.com

ILLUSTRATED HISTORY
Ultimate Triumph Collection	$49.95
American Police Motorcycles - Revised	$24.95

GUIDE BOOKS
Honda Motorcycles - Enthusiast Guide - 1959-1985	$27.95

BIKER BASICS
Custom Bike Building Basics	$24.95
Sportster/Buell Engine Hop-Up Guide	$24.95
Sheet Metal Fabrication Basics	$24.95
How to Fix American T-Twin Motorcycles	$27.95

COMPOSITE GARAGE
Composite Materials Handbook #1	$27.95
Composite Materials Handbook #2	$27.95
Composite Materials Handbook #3	$27.95

HOT ROD BASICS
So-Cal Speed Shop's How to Build Hot Rod Chassis	$24.95
Hot Rod Wiring	$27.95
How to Chop Tops	$24.95

CUSTOM BUILDER SERIES
How to Build A Café Racer	$27.95
Advanced Custom Motorcycle Wiring - Revised	$27.95
How to Build an Old Skool Bobber Sec Ed	$27.95
How To Build The Ultimate V-Twin Motorcycle	$24.95
Advanced Custom Motorcycle Assembly & Fabrication	$27.95
How to Build a Cheap Chopper	$27.95

MOTORCYCLE RESTORATION SERIES
Triumph Restoration - Unit 650cc	$29.95
Triumph MC Restoration Pre-Unit	$29.95

SHEET METAL
Advanced Sheet Metal Fabrication	$27.95
Ultimate Sheet Metal Fabrication	$24.95
Sheet Metal Bible	$29.95

AIR SKOOL SKILLS
Airbrush Bible	$29.95
How Airbrushes Work	$24.95

PAINT EXPERT
How To Airbrush, Pinstripe & Goldleaf	$27.95
Kosmoski's New Kustom Painting Secrets	$27.95
Pro Pinstripe Techniques	$27.95
Advanced Pinstripe Art	$27.95

TATTOO U Series
Advanced Tattoo Art - Revised	$27.95
Cultura Tattoo Sketchbook	$32.95
Jim Watson's Tattoo Sketchbook	$32.95
Into The Skin The Ultimate Tattoo Sourcebook	$34.95
American Tattoos	$27.95
Tattoo Bible Book One	$27.95
Tattoo Bible Book Two	$27.95
Tattoo Bible Book Three	$27.95

LIFESTYLE
Bean're — Motorcycle Nomad	$18.95
George The Painter	$18.95
The Colorful World of Tattoo Models	$34.95

Resources

Baileigh Industrial Tools
www.baileighindustrial.com

Justin Baker
www.justinbaker.co.uk

Mitch Bell
abzperformance@gmail.com

Neil Dunder
www.gogitzit.com

Amber Burchette
Amber-Burchette-Racing/Facebook

Jamie Downie
www.kustomgarage.com.au

Randy Ferguson
www.FergusonCoachbuilding.com

Ron Fournier
www.FournierEnterprises.com

Dagger Tools
www.daggertools.com

David Gardiner
www.metalshapingzone.com

Gulley Tools
www.gulleyperformancecenter.com

Hoosier Profiles
www.hoosierprofiles.com

Lazze Jansson
www.lazzemetalshaping.com

Metal Ace
www.englishwheels.net

Geoff Moss
www.mphmotorpanels.com

The Panel Shop
www.thepanelshop.net

Kerry Pinkerton
www.wheelingmachines.com

Wray Schelin
www.proshaper.com

Peter Tommasini
www.handbuilt.net.au

Kent White
www.tinmantech.com

Furthering Your Education

Reading books, and watching videos are a good way to learn the vocabulary and fundamentals of metalshaping, but attending classes and workshops are much better. Classes are offered around the world, and their quality and content varies greatly, as does cost. I'm listing classes and videos that I can personally recommend. Please do internet searches for their current contact information.

Jamie Downie, *Kustom Garage*, Melbourne, Australia. Jamie is not one to waste time. His instruction is "on task", personable, and flexible to your needs. Jamie is widely experienced in cars and bikes, and he offers Yoder power hammer instruction.

Randy Ferguson, *Ferguson Coachbuilding*, Robinson, Illinois. Randy routinely gives e-wheeling lessons at Metal Meets, but is also available for individualized courses. An excellent, patient teacher with a well-equipped shop.

Ron Fournier, *Fournier Enterprises*, Shelby Township, Michigan. Ron's three-day class has gotten many metalshapers started. A true gentleman and great raconteur. His videos are particularly good.

David Gardner, *Classic Metal Shaping*, Harwich, UK. If you only buy one instructional video in your life, get David's *Bodywork Restoration Tutorial,* a monumental thesis on the art. He makes it look so easy, and with few specialized tools. *Highest recommendation!*

Geoff Moss, *MPH Motorpanels*, Liskeard, UK. If you want to learn traditional English panel beating techniques then Geoff is your man. His week long course will get you up and wheeling fast. Excellent reputation throughout the UK.

Wray Schelin, *Pro Shapers*, Stafford Springs, Connecticut. Wray adapts his classes to suit your skill level and gives you plenty of time to work on them after class hours, many repeat students. Highly recommended! Excellent video on the safe and smooth shrinking disc, as well as tools available.

Peter Tommasini, *Handbuilt*, Melbourne, Australia. Peter offers great classes both at his shop in Melbourne, but also around Australia. His instruction is fast, friendly, and suitable for all skill levels. His video collection is a "must have" for serious learners, especially the Monaro quarter panel set. In addition to classes and videos, he offers an excellent range of specialist tools. Highly recommended.

Kent White, *T-M Technologies*, various locations, US and Europe. Kent's classes include hands on instruction, and can be geared to specific industrial or corporate needs. Kent is the only one on this list who can provide airworthiness instruction. Very popular with attendees and profoundly informative. His video collection is extremely good. Outstanding line of high quality and industrial grade tools available.

www.ingramcontent.com/pod-product-compliance
Lightning Source LLC
Chambersburg PA
CBHW081422230426
43668CB00016B/2316